Synthesis Lectures on Engineering, Science, and Technology

The focus of this series is general topics, and applications about, and for, engineers and scientists on a wide array of applications, methods and advances. Most titles cover subjects such as professional development, education, and study skills, as well as basic introductory undergraduate material and other topics appropriate for a broader and less technical audience.

Supriyo Bandyopadhyay

Magnetic Straintronics

An Energy-Efficient Hardware Paradigm
for Digital and Analog Information
Processing

 Springer

Supriyo Bandyopadhyay ⓘD
Department of Electrical and Computer
Engineering
Virginia Commonwealth University
Richmond, VA, USA

ISSN 2690-0300   ISSN 2690-0327 (electronic)
Synthesis Lectures on Engineering, Science, and Technology
ISBN 978-3-031-20685-6   ISBN 978-3-031-20683-2 (eBook)
https://doi.org/10.1007/978-3-031-20683-2

This Springer imprint is published by the registered company Springer Nature Switzerland AG
The registered company address is: Gewerbestrasse 11, 6330 Cham, Switzerland

*Dedicated to Bela Bandyopadhyay,
Anuradha Bandyopadhyay and
Saumil Bandyopadhyay*

Preface

This monograph introduces the readers to some of the most important topical areas of *magnetic straintronics* that could lead to a potent hardware platform for processing digital and analog information at a minimal energy cost. Magnetic straintronics, hereafter referred to simply as "straintronics" for the sake of brevity, is the rapidly burgeoning field of science and engineering dealing with manipulating the magnetic states of magnetostrictive nanomagnets with electrically generated mechanical strain. These nanomagnets can implement a wide range of functional devices and circuits that are extremely energy-efficient and hence ideal for many applications, such as embedded processors, medically implanted devices, aggressively miniaturized stealth devices for defense and crime-fighting, and even mainframe computers.

The primitive unit of all digital hardware (computers, signal processors, communication devices, etc.) is a *binary switch* that has two "states" encoding the binary bits 0 and 1. Digital computing and signal processing are carried out by switching between the two states. The best switch is the one that can switch the fastest with minimal energy dissipation while boasting a low switching error rate for reliability. A straintronic switch, fashioned out of a magnetostrictive nanomagnet whose magnetization can assume only two stable orientations (representing bits 0 and 1), may be able to switch in less than 1 ns with energy dissipation as low as ~1 aJ when the magnetization is switched with electrically generated mechanical strain. The associated energy-delay product is one order of magnitude lower than what the best silicon field effect transistors today can offer. The low energy-delay product, of course, comes at a cost, and the cost is usually a high switching error rate. That high rate is a serious disadvantage for some applications, such as Boolean logic, but not so much for collective computational models (e.g. neuromorphic networks) where the collective activities of many devices working cooperatively elicit the computational or signal processing activity, and the failure of one or few devices to act properly does not impair the circuit operation. Straintronic devices have other attractive attributes as well, such as unusual characteristics that can be exploited to build better circuits, and, of course, non-volatility, which can be leveraged to implement superior architectures for specific digital and analog circuits. Non-volatility implies that the information in the switches (i.e. which of the two states it is in) is not lost once the device is powered off.

Non-volatility can offer other advantages as well, such as resilience to hacking, energy saving (accruing from the fact that data do not have to be refreshed periodically), and no boot delay when powering up a computer.

This monograph mostly (but not exclusively) presents research carried out in the author's group and in the groups of his collaborators in the area of magnetic straintronics. The author acknowledges all his students who have worked in the field (too many to name individually) and his collaborators who (in alphabetical order) are Profs. Jayasimha Atulasimha, Anjan Barman, Avik W. Ghosh, Bin Ma, Csaba Andras Moritz, Erdem Topsakal, Amit Ranjan Trivedi and Jian-ping Wang.

Some of this monograph is excerpted from the following two review articles co-authored by this author:

1. S. Bandyopadhyay, J. Atulasimha and A. Barman, Magnetic straintronics: manipulating the magnetization of magnetostrictive nanomagnets with strain for energy-efficient applications. Appl. Phys. Rev. **8**(4), 041323 (2021)
2. N. D'Souza, et al., Energy-efficient switching of nanomagnets for computing: straintronics and other methodologies. Nanotechnology **29**(44), 442001 (2018)

Comments on this work are welcome and can be emailed to the author at sbandy@vcu.edu.

Richmond, VA, USA   Supriyo Bandyopadhyay

Contents

Magnetic Straintronics

1

1.1   Introduction: Energy and Information

The two most important infrastructures in our society are energy and information because they define our civilization's level of technical advancement. The Kardashev scale [1] measures a civilization's technical prowess in terms of a simple metric: a type I civilization is able to harness all the power that reaches its home planet from the parent star, which in our case, is about 2×10^{17} watts. A type II civilization would be able to harness all the power radiated by its parent star, including that which does not reach its planet. In our case, this will be 4×10^{26} watts. A Type III civilization, on the other hand, can harness all the power in its galaxy. For the Milky Way, this is about 4×10^{37} watts. Based on these power harvesting capabilities, a Type II civilization will be able to send the information contained in a medium sized library across intergalactic distances of ten million light years after transmitting for several weeks, and a Type III civilization can accomplish that same feat after transmitting for a mere 3 s [2].

It may not be pure happenstance that Kardashev thought of a civilization using its energy resources for information transmission rather than any other task. *It does take considerable amount of energy to store, process and communicate information.* Curiously, Carl Sagan had added another dimension to the Kardashev scale to define a civilization's state of progress and it directly concerns information [3]. He assigned the letter A to represent 10^6 unique bits of information that a civilization possesses and can process, and each successive letter in the alphabet to represent an order of magnitude increase so that a level Z civilization would have 10^{31} bits of information at its disposal. In 2018, we were a level 0.73 J civilization. All this suggests that there is an intrinsic and unavoidable link between the energy available to a civilization and its information processing and transmitting capability.

S. Bandyopadhyay, *Magnetic Straintronics*, Synthesis Lectures on Engineering, Science, and Technology, https://doi.org/10.1007/978-3-031-20683-2_1

Why is this the case? The fact of the matter is that information is "physical" [4] and therefore it takes *energy* to handle or communicate *information*. Today, roughly 10% of the energy produced in the United States is consumed by all manner of information processing devices (computers, cell phones, health monitoring devices) and it will likely grow to 25% by the year 2030. A modern data center can require as much energy as a major metropolis and its carbon footprint can exceed that of a small nation. Prior to the war with Russia, Ukraine had announced a plan to build a data center for mining cryptocurrency data next to a nuclear power plant because the power requirement was anticipated to be 2–3 GW [5]! We, humans, are information processors as well. We constantly process information received from the five senses and use roughly 20% of the calories that we consume to "think" and process that information. Since information processing is so demanding of energy, the search is always on in our society to find increasingly more energy-efficient devices and hardware for information processing.

Conventional information processors used in modern everyday life, such as computers, employ *electric charge-based devices*, e.g. field-effect transistors, to implement the hardware for carrying out computational tasks. These charge-based devices have two innate drawbacks. First, they are not frugal in their use of energy when they process digital information, say, by switching between two states encoding the binary bits 0 and 1 (more on this later), and second, they are "volatile" and cannot retain information for long (since charges leak out), so that the information content has to be refreshed periodically at considerable energy cost. This has motivated the search for alternate state vectors—other than charge—to encode information. One important alternate is the quantum mechanical spin of an electron which is at the core of magnetism. Magnetic devices are usually non-volatile, unlike charge-based devices, and hence this is an enticing alternative. Bistable magnetic devices, whose magnetization has two stable orientations, can replace field-effect transistors as binary switches. Their two stable magnetization orientations can encode the binary bits 0 and 1. Switching between them can be accomplished in a very energy-efficient way *sometimes* (depending on how the switching is carried out). Finally, the non-volatility can be leveraged to build superior hardware architectures that can produce systems with lower energy consumption, smaller circuit footprint, and faster processing speed. Additionally, the non-volatility can enable large scale edge computing, where almost all the information processing is carried out locally (in the edge device) without the need to access the cloud. That offers resilience against cloud-borne cyberattacks.

Unfortunately, not all magnet switching mechanisms are energy-efficient. Most are not and they will dissipate orders of magnitude more energy than what a field-effect transistor will dissipate to switch, which begs the question if they are worthy of consideration despite the attractive proposition of non-volatility. Opinion is divided on this issue and we will not discuss it here. There is, however, at least one magnet switching mechanism which will dissipate less energy than what a state of the art transistor dissipates to switch, and that is the switching of magnetization with electrically generated mechanical strain (straintronics). This monograph discusses *straintronic* devices and their applications. It is a field that has seen explosive growth over the last decade because of its promise of providing a non-volatile, extremely energy-efficient and cyber-secure platform for all manner of digital and analog information processing.

Straintronic devices that we will discuss in this monograph are usually made of a "two-phase multiferroic" comprising a magnetostrictive magnet delineated on (and elastically coupled to) an underlying piezoelectric film. Application of a voltage across the piezoelectric film with a suitable configuration of electrodes generates a particular distribution of mechanical strain in the piezoelectric which is transferred wholly or partially to the magnet. That changes the latter's magnetization state (e.g. the orientation of the magnetization) via the inverse magnetostriction, or the Villari effect [6], sometimes also called the magneto-elastic effect. The switching can be completed in less than 1 ns and the energy dissipated in the switching action can be as low as ~1 aJ, which is about two orders of magnitude lower than the energy dissipated in switching a state-of-the-art field-effect transistor, and several orders of magnitude lower than the energy dissipated in most other magnet switching methodologies, such as switching with an on-chip local magnetic field or with a spin current passed through the magnet.

Unfortunately, nothing comes without a cost and the exceptional energy efficiency of straintronics must also come at a cost. There is always a trade-off between energy efficiency and error resilience [7]; devices that consume less energy to work are also usually less reliable. The Achilles' heel of straintronic devices is their relatively high switching error rate which degrades their reliability. Fortunately, this is not debilitating in many applications, such as analog applications, non-Boolean logic applications, neuromorphic networks, probabilistic computing, and collective computation, where the cooperative action of many straintronic devices, acting in unison, elicit the computational activity and the failure of a single device, or even a small subset of all devices, does not impair the circuit functionality.

The following chapters describe some of the unique and fascinating attributes of straintronic devices which can be leveraged for myriad applications and offer not just energy efficiency, but also reduced computational requirements and sometimes reduced device footprint, not to mention non-volatility.

References

1. N.S. Kardashev, On the inevitability and possible structures of super civilizations. https://articles. adsabs.harvard.edu/pdf/1985IAUS..112..497K
2. N.S. Kardashev, Transmission of information by extra-terrestrial civilizations. Sov. Astron. –AJ, **8**, 217 (1964)
3. C. Sagan, *Carl Sagan's Cosmic Connection: An Extraterrestrial Perspective* (Cambridge University Press, Cambridge, UK, 2000)
4. R. Landauer, Information is physical. Phys. Today **44**(5), 23 (1991)

5. https://www.datacenterdynamics.com/en/news/ukraine-plans-huge-cryptocurrency-mining-data-centers-next-nuclear-power-plants/
6. https://encyclopedia2.thefreedictionary.com/Villari+Effect
7. R. Rahman, S. Bandyopadhyay, The cost of energy-efficiency in digital hardware: the trade-off between energy dissipation, energy–delay product and reliability in electronic, magnetic and optical binary switches. Appl. Sci. **11**, 5590 (2021)

Binary Switches for Digital Information Processing

2

Although straintronics has many applications, the one that has attracted most attention is understandably binary switches since it is the primitive unit of all digital information processing hardware. A binary switch has two states—ON and OFF—which encode the binary bits 0 and 1. The quintessential binary switch and the one used to benchmark all digital switches, is the "metal–oxide–semiconductor-field-effect-transistor" (MOSFET) and its various incarnations such as the "fin field effect transistor" (FINFET), "tunnel-field-effect-transistor" (TFET), "negative capacitance transistor" (n-CFET), etc. It has two conductance states—high (ON) and low (OFF)—which encode the binary bits. Energy is dissipated when the transistor switches between these two states.

2.1 Charge-Based Switches: The Transistor

The transistor is the dominant charge based switching device that rules the roost in digital electronic circuits. Because it is a charge based device, it is volatile. It is also not particularly energy-efficient since it dissipates a significant amount of energy when it switches. One can get an estimate of this energy from available specifications of high-performance processors currently in the market. The AMD Ryzen 9 5900 X is a modern processor built with 7-nm transistor technology and was released late in the year 2020. The chip has 4.15 billion transistors, operates with a power supply voltage of 1.5 V, dissipates a power of 105 watts and operates at a clock frequency of 3.7 GHz [1]. We can assume that 10% of the transistors in the chip switch at any given time and that almost all of the energy dissipated in the chip comes from the switching of the transistors. This yields an estimate for the energy dissipated on the average by a transistor during a single switching event. That quantity is

$$E_d = P_d / (Naf) = 105 / \left(4.15 \times 10^9 \times 0.1 \times 3.7 \times 10^9\right) = 68.3 \text{ aJ}, \tag{2.1}$$

S. Bandyopadhyay, *Magnetic Straintronics*, Synthesis Lectures on Engineering, Science, and Technology, https://doi.org/10.1007/978-3-031-20683-2_2

where P_d is the power dissipated by the chip, N is the number of transistors, a is the fraction of the transistors switching on the average at any given time, and f is the clock frequency. This estimate is about one-half of the energy dissipation estimated for the Intel® Core™ i7-6700 K processor built with 14 nm transistor technology and released in the year 2015 [2]. Five years after the release of the Intel® Core™ i7-6700 K processor and a two-fold reduction in transistor feature size, the energy dissipated per transistor per switching event decreased only by a factor of 2. This highlights the difficulty of reducing energy dissipation in a transistor.

This somber realization behooves us to look at the origin of the energy dissipation in a transistor, i.e. where does it come from? The basic structure of a MOSFET is shown in Fig. 2.1. The transistor is turned ON when a gate voltage draws mobile charges (of polarity opposite to that of the gate charges) into the channel from the source under Coulomb attraction. This establishes a conducting path between the source and the drain and turns the transistor ON. The transistor turns OFF when the gate voltage (and the gate charges) reverse polarity which expels the mobile charges from the channel because of Coulomb repulsion, thereby disrupting the conducting path between the source and the drain. The switching action is therefore associated with charges flowing in and out of the channel.

The point to note is that the two states ON and OFF, encoding the binary bits 0 and 1, are demarcated by the amount of electrical charge in the channel. A larger amount of channel charge Q_1 (which establishes a conducting path between the source and the drain)

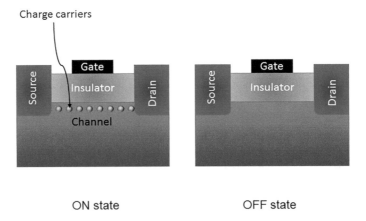

ON state   OFF state

Fig. 2.1 The working principle of an enhancement-mode MOSFET transistor with source, gate and drain terminals. The transistor is on when the charges placed on the gate by a gate voltage draws mobile charges of opposite polarity from the source contact into the channel underneath the insulator separating the gate from the semiconductor channel. These mobile charges establish a conducting path between the source and the drain, allowing flow of a current between these two contacts, thereby turning the transistor ON. When the gate voltage reverses polarity, the mobile charges in the channel are expelled from the channel owing to Coulomb repulsion and the transistor turns OFF

represents the ON-state and a smaller amount Q_2 (which disrupts the conducting path) represents the OFF-state. To switch the device, one must change the amount of charge from Q_1 to Q_2, or vice versa, thereby causing the flow of a (time-averaged) current I given by.

$$I = |Q_1 - Q_2|/\Delta t, \tag{2.2}$$

where Δt is the amount of time it takes to change the channel charge from Q_1 to Q_2, or vice versa. The energy dissipated in this action is

$$E_d = I^2 R \Delta t = \left(\Delta Q/\Delta t\right) I R \Delta t = \Delta Q I R = \Delta Q \Delta V. \tag{2.3}$$

where R is the resistance in the path of the current and $\Delta V = I R$. We can think of ΔV as the amount of voltage needed to induce the current I through the resistance R and thereby change the charge in the channel by the amount ΔQ [2–4].

Contrary to appearances, the energy dissipation given in Eq. (2.2) is actually *not* independent of the switching time Δt (even though Δt does not explicitly appear in the expression), because ΔV depends on the switching time for a fixed ΔQ and R $\left(\Delta V = \Delta Q R/\Delta t\right)$. If we re-write Eq. (2.3) as $E_d = (\Delta Q)^2 R/\Delta t$, we immediately see that the energy dissipation is inversely proportional to the switching time, meaning that we will dissipate *more* energy if we switch *faster* (smaller Δt). Speed therefore comes at the cost of energy and vice versa. Hence, a more meaningful quantity may be the "energy-delay product" defined as $E_d \Delta t = (\Delta Q)^2 R$. This is a quantity frequently used to benchmark a switching device, although it may not be the best benchmarking index since, as we have stated in the previous chapter, we can always buy energy efficiency at the cost of reliability. A better benchmarking index will be quantity that combines the energy delay product with the switching error rate. This important issue motivates the ensuing discussion.

One cannot reduce the energy-delay product $E_d = (\Delta Q)^2 R/\Delta t$ arbitrarily by reducing ΔQ, since a sufficiently large ΔQ is needed to distinguish between bits 0 and 1. If one reduces ΔQ, the two bits will become increasingly indistinguishable, which would rapidly reduce discernibility of the bits when operating in a noisy environment. This will increase the error probability in information processing. One cannot reduce R arbitrarily either, since that would require one to increase the cross-section of the current path at the cost of a larger device footprint. All this hints at the fact that there must be a lower bound on the energy that we have to dissipate if we use a charge-based device like the transistor as a binary switch [3, 4]. This bound will be established by the minimum reliability that we can tolerate.

2.1.1 Minimum Energy Dissipation in Charge-Based Switches for Reliability

From Eq. (2.3), we can get an estimate of the amount of charge that is moved into and out of the channel of a modern-day MOSFET (e.g. in the AMD Ryzen 9 5900 X processor) to switch it from ON to OFF, or vice versa. Since the power supply voltage is 1.5 V in the AMD processor, $\Delta V = 1.5$ V. Hence, the charge moved into and out of the channel is

$$\Delta Q = E_d \big/ \Delta V = 68.3 \times 10^{-18} \big/ 1.5 = 4.553 \times 10^{-17} \text{ Coulombs,} \qquad (2.4)$$

which is the charge carried by only 283 electrons or holes in the semiconductor channel. This is a very small number of charge carriers.

The number of charge carriers that spontaneously appear/disappear in the channel due to noise and thermal fluctuations has to be much smaller than this number so that the transistor does not randomly switch on and off because of noise. This latter number is related to the charge fluctuation in the transistor's "gate" and is given by [5]

$$\Delta Q|_{\text{fluctuation}} = \sqrt{C_g kT}, \qquad (2.5)$$

where C_g is the gate capacitance, k is the Boltzmann constant and T is the absolute temperature. The gate capacitance for the MOSFET used in the AMD processor can be estimated roughly as (this includes contributions of line capacitance, stray capacitance, etc.)

$$C_g = \Delta Q \big/ \Delta V = 4.553 \times 10^{-17} \big/ 1.5 = 30 \text{ aF,} \qquad (2.6)$$

which makes $\Delta Q|_{\text{fluctuation}} = 3.56 \times 10^{-19}$ Coulombs at room temperature, and that is the charge of only ~2 charge carriers. Thus, ~2 charge carriers can spontaneously appear or disappear in the channel of the AMD MOSFET at room temperature. Fortunately, that number is only 0.7% of the charge carriers that must flow into and out of the channel to make the MOSFET switch. The 0.7% variation is not a large enough variation to make the transistor unreliable and hence the Ryzen processor is not error-prone.

We must ensure that the random charge that appears in the channel is always a small fraction of the charge that is required to be moved around to switch the transistor, lest the switching action becomes random and unreliable. This will set a lower limit on the amount of charge that needs to be moved into or out of the channel to switch the transistor. Let us call it $\Delta Q_{\min} = \xi \, \Delta Q|_{\text{fluctuation}}$, where we can interpret $\xi(\xi \gg 1)$ as a measure of the reliability of switching. In that case, the minimum energy that we must dissipate to switch a FINFEET/MOSFET type device will be [4]

$$E_d^{\min} = \Delta Q_{\min} \Delta V_{\min} = (\Delta Q)^2 \big/ C_g = (\xi \, \Delta Q|_{\text{fluctuation}})^2 \big/ C_g = \xi^2 kT. \qquad (2.7)$$

Equation (2.7) tells us that the minimum energy that one must dissipate to switch a MOSFET (or its various clones) is determined by the minimum reliability (or minimum ξ) that one is able to tolerate. There is a clearly a trade-off between energy dissipation and reliability that is obvious from Eq. (2.7); one can always sacrifice reliability for energy dissipation and vice versa, as we had already mentioned in the previous chapter, and Eq. (2.7) illustrates that nicely.

Let us say that we do not want the random charge fluctuation in the channel to be more than 10% of the charge that we have to move around to switch. That is our standard for reliability. In that case, we want $\xi > 10$, and hence the minimum energy dissipation that we will have to live with will be ~100 kT, which is 0.42 aJ if we are operating at room temperature. This is more than two orders of magnitude smaller than what we are currently dissipating in the Ryzen transistor, so there is still plenty of room for improvement. It is instructive to point out that the minimum dissipation in a transistor with ξ = 10 is ~145 times larger than the Landauer limit of $kTln(2)$ for a binary switch that is switched irreversibly [6]. If one were to operate the MOSFET at the Landauer limit, ξ would be 0.83 which would actually make the switch very unreliable.

The minimum energy-delay product, on the other hand, is given by

$$(E_d \Delta t)^{min} = (\Delta Q_{min})^2 R = \xi^2 kT R C_g. \tag{2.8}$$

We cannot reduce this quantity arbitrarily either, since reducing the gate capacitance C_g will reduce the transistor's transconductance and reducing the access resistance R will increase the device footprint. In the end, there are always certain technological impasses that cannot be overcome.

2.2 Magnetic Switches

A bistable magnetic "switch" can be implemented simply with a nanoscale ferromagnet shaped like an elliptical disk as shown in Fig. 2.1. If its dimensions are sufficiently small (but not so small as to make it super-paramagnetic instead of ferromagnetic at the temperature of operation) and its eccentricity is also not too small, then the nanomagnet will contain a *single* magnetic domain wherein all the spins in the nanomagnet will point in the same direction owing to exchange interaction between them. Additionally, if the ferromagnet is amorphous or polycrystalline, so that it does not possess any magneto-crystalline anisotropy, then the magnetization can point only right or left along the major axis of the ellipse as shown in Fig. 2.1a, which will make the magnetization "bistable". These two stable orientations of the magnetization can then be used to encode the binary bits 0 and 1. Switching the magnetic switch would involve simply flipping the magnetization from right to left, or vice versa, with some agent, which need not require passing current through the nanomagnet (although some current may have to be passed in some external peripheral element to make the flipping take place).

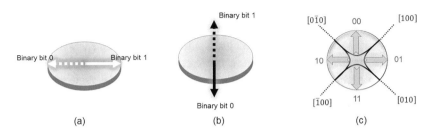

Fig. 2.2 **a** A single-domain amorphous nanomagnet shaped like an elliptical disk and possessing in-plane magnetic anisotropy has two stable magnetization orientations which point along the major axis, either to the left or to the right, **b** a single-domain amorphous nanomagnet shaped like an elliptical or circular disk and possessing perpendicular magnetic anisotropy has two stable magnetization orientations which are perpendicular to the surface, either pointing up or down, and **c** a single domain crystalline nanomagnet possessing magneto-crystalline anisotropy can have multiple stable orientations along different crystallographic directions along the plane (in this case the (100) plane), which will allow encoding more than two bits in the magnetization orientations

The type of nanomagnet shown in Fig. 2.2a is said to possess *in-plane* (magnetic) anisotropy (IPA). If the thickness of the nanomagnet is small enough, then the magnetization can point perpendicular to the surface (not along the major axis) owing to the large surface anisotropy. In this case, the two stable orientations are "up" and "down", as shown in Fig. 2.2b. Such a nanomagnet is said to possess perpendicular magnetic anisotropy (PMA). The magnetization states of PMA nanomagnets are relatively insensitive to imperfections such as edge roughness and these nanomagnets are more scalable in size, i.e. their lateral dimensions can be made smaller without causing them to lose their ferromagnetism. They can be either circular or elliptical in shape and are preferred in many applications, mostly memory related applications where scalability is important. The IPA nanomagnets, on the other hand, may be preferable in some other applications such as in probabilistic computing because they respond faster in those applications.

If the nanomagnet is crystalline and has magneto-crystalline anisotropy, then more than two stable states are possible as shown in Fig. 2.2. The nanomagnet does not have to be elliptical in this case, and could be circular, since the stable orientations accrue from magneto-crystalline anisotropy and not shape anisotropy. The stable orientations are along different crystallographic directions. In this case, the nanomagnet is multistable and can be used to encode more than two bits, which can then allow implementing such constructs as four-state logic [7].

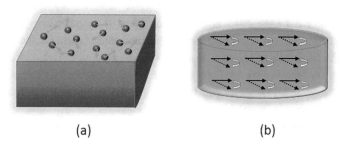

Fig. 2.3 **a** Uncorrelated motion of information carriers (electrons or holes) in a transistor channel, and **b** correlated motion of information carriers (spins) in a single domain nanomagnet

2.3   Correlated Motion in Single Domain Magnetic Switches

A very important difference between the transistor switch and the nanomagnet switch is that in the transistor channel, the motions of the individual charge carriers which carry the information about the state of the device (ON or OFF) are *uncorrelated*. In a single domain nanomagnet, on the other hand, when the magnetization rotates from one direction to another, *all* the individual spins, which are the information carriers, rotate together in unison, with perfect coordination, because the exchange coupling tends to keep them always parallel to each other. This is shown in Fig. 2.3. As a result, all the spins together behave as *one* giant classical spin [8]. Thus, the motions of information carriers are *correlated* in a nanomagnet, but not in a transistor. If there are N electrons or holes (information carriers) in a transistor channel, then they have N degrees of freedom because of the lack of correlation, whereas if there are M spins (information carriers) in a nanomagnet, they collectively have a *single* degree of freedom. Consequently, the *minimum* energy that must be dissipated in switching a transistor cannot be any less than $NkTln(1/p)$ [where p is the switching error probability] whereas in a nanomagnet, it cannot be any less than $kTln(1/p)$ [9]. This can give the nanomagnet an intrinsic advantage over the transistor, but it is rather academic, since the energy we actually dissipate in switching either entity is currently much larger than these limits.

2.4   Reading the Bit State of a Magnetic Switch by Electrical Means: Spin to Charge Conversion

The bit state of a magnetic switch, i.e. whether it is in the state representing bit 0 or bit 1, is read electrically by a construct called a *magnetic tunnel junction* (MTJ). It is a three layered device as shown in Fig. 2.4. There is a hard (or "fixed") ferromagnetic layer whose magnetization orientation is fixed in one direction and a soft (or "free") ferromagnetic layer whose magnetization orientation is bistable and can point in one of two stable directions representing the binary bits 0 and 1. One of these directions is *parallel* to the magnetization of the hard layer and the other is *antiparallel*. The resistance of the MTJ, measured between the two ferromagnetic layers, is low in the "parallel" state and high in the "antiparallel" state because of spin-dependent tunneling between the two layers through the intervening insulating spacer layer. Therefore, by measuring the electrical resistance of the MTJ, one can tell whether the magnetization of the soft layer, hosting the binary bit, is parallel or antiparallel to that of the hard layer. Since the hard layer's magnetization is fixed and known apriori, one can therefore tell in which direction the soft layer's magnetization is pointing, and hence whether its magnetization state is encoding bit 0 or 1. Because this construct converts magnetic information (magnetization orientation) into electrical information (electrical resistance), it is sometimes referred to as *spin-to-charge conversion*. The conversion process is energy-expensive since it involves the flow of current through the MTJ. Many magnetic circuit proposals, especially those purporting to implement Boolean logic, strive to avoid spin-to-charge conversion at intermediate stages (e.g. between two successive logic gates) and do the conversion only at the final stage in order to reduce energy dissipation [10].

Two major differences between an MTJ and a field-effect transistor—both of which have two conductance states: "high" and "low"—are that ratio of the two conductance states is typically $\sim 10^5$ in the case of the transistor but less than 10 in the case of the MTJ at room temperature, and that the conductance state is *volatile* in the case of the transistor but *non-volatile* in the case of the MTJ. These differences play a crucial role in the design of digital circuits with MTJs and transistors.

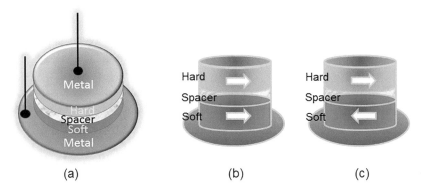

(a)   (b)   (c)

Fig. 2.4 **a** Basic three-layered structure of a magnetic tunnel junction (MTJ), **b** the low-resistance state where the magnetization orientations of the hard and soft layer are mutually parallel, and **c** the high-resistance state where they are mutually antiparallel

References

1. https://www.techpowerup.com/cpu-specs/ryzen-9-5900x.c2363
2. S. Datta, V.-Q. Diep, B. Behin-Aein, What constitutes a nanoswitch? A perspective, in *Emerging Nanoelectronic Devices*, ed. By A. Chen, J. Hutchby, V. Zhirnov, G. Bourianoff (Wiley, Chichester, UK, 2015), pp. 15–34
3. S. Bandyopadhyay, J. Atulasimha, A. Barman, Magnetic straintronics; manipulating the magnetization of magnetostrictive nanomagnets with strain for energy-efficient applications. Appl. Phys. Rev. **8**, 041323 (2021)
4. S. Bandyopadhyay, Straintronics: digital and analog electronics with strain switched nanomagnets. IEEE Open J. Nanotechnol. **1**, 57 (2020)
5. L.B. Kish, End of Moore's law: thermal (noise) death of integration in micro and nano electronics. Phys. Lett. A **305**, 144 (2002)
6. R. Landauer, Irreversibility and heat generation in the computing process. IBM J. Res. Develop. **5**, 183 (1961)
7. N. D'Souza, J. Atulasimha, S. Bandyopadhyay, Four state nanomagnetic logic using multiferroics. J. Phys. D **44**, 265001 (2011)
8. R.P. Cowburn, D.K. Koltsov, A.O. Adeyeye, M.E. Welland, D.M. Tricker, Single domain circular nanomagnets. Phys. Rev. Lett. **83**, 1042 (1999)
9. S. Salahuddin, S. Datta, Interacting systems for self-correcting low power switching. Appl. Phys. Lett. **90**, 093503 (2007)
10. B. Behin-Aein, D. Datta, S. Salahuddin, S. Datta, Proposal for an all-spin logic device with built-in memory. Nature Nanotech **5**, 266–270 (2010)

Switching a Magnetic Switch with an Electrical Current or Voltage

3

A nanomagnet's magnetization can be switched from one stable orientation to another in many ways. Straintronics is one of the most energy-efficient approaches, but there are many others with varying degrees of energy efficiency. In general, one can switch the magnetization electrically using either a current flowing through some element (not necessarily the nanomagnet itself) or a voltage applied across some element (again, not necessarily the nanomagnet itself).

3.1 Current-Controlled Mechanisms for Switching the Magnetization of Nanomagnets

3.1.1 Switching with a Local Magnetic Field Generated by a Current

An on-chip current (not flowing through the nanomagnet) can generate an on-chip magnetic field which can switch the magnetization orientation of a nanomagnet on the chip to align along the direction of the field. This approach has been used in the past, but turns out to be extremely energy-inefficient since it takes a lot of current to generate the required magnetic field [1]. An additional drawback is that it is very difficult to confine a magnetic field to a small area, which makes it challenging to address a single nanomagnet in a dense array. Therefore, the array density has to be reduced significantly in order to switch with a current generated magnetic field, which is not desirable.

© The Author(s), under exclusive license to Springer Nature Switzerland AG 2022
S. Bandyopadhyay, *Magnetic Straintronics*, Synthesis Lectures on Engineering, Science, and Technology, https://doi.org/10.1007/978-3-031-20683-2_3

3.1.2 Spin-Transfer Torque

A better option is to *inject* a spin polarized current into a nanomagnet using a spin polarizer which is usually another magnet. This allows one to address a nanomagnet individually, even in a dense array. The injected spins transfer their angular momenta to the resident spins in the nanomagnet and make the latter's magnetization align in the direction of the injected current's spin polarization. This switches the magnetization to the desired orientation. It is also possible to *extract* (as opposed to inject) a spin polarized current from a nanomagnet with a spin analyzer (by simply reversing the polarity of the current that passes through the nanomagnet) which depletes the nanomagnet of the population of spins that are present in the extracted current and therefore makes the nanomagnet's magnetization align in the direction opposite to the spin polarization of the extracted current. Thus, a bistable nanomagnet's magnetization can be switched to either stable orientation by injecting or extracting a spin polarized current. These mechanism is referred to as "spin transfer torque" (STT) [2] since the injected spins transfer their angular momentum to the resident spins in the nanomagnet and in the process exert a torque on them, which then produces a rotating torque on the nanomagnet's magnetization. This is a popular approach that is currently used in *spin-transfer-torque-random-access-memory* (STT-RAM) to orient the magnetization of a nanomagnet and thus write bits into non-volatile memory made of nanomagnets. A slight variation of this strategy employs the giant spin Hall effect (GSHE) in some heavy metals (e.g., Pt or β-Ta) [3] or spin-momentum locking in topological insulators (e.g., Bi_2Se_3) [4]. These materials have a special property. When a charge current is injected into a slab of such a material, the two opposite surfaces become oppositely spin polarized resulting in the flow of a spin current perpendicular to the surface of the slab. If a nanomagnet is placed on one of the surfaces, a spin current will diffuse into the nanomagnet (because that surface is spin polarized) and align the magnetization of the nanomagnet in the direction of the diffusing spins. We can flip the magnetization back to the opposite direction by simply reversing the direction of the charge current. These mechanisms are illustrated in Fig. 3.1.

3.2 Spin–orbit Torque

There are other current induced mechanisms for switching the magnetizations of nanomagnets, one of which is called *spin-orbit torque* (SOT) [5]. Here, the magnetization of a nanomagnet with perpendicular magnetic anisotropy (PMA) is switched with an in-plane current in an asymmetric structure with Rashba spin-orbit interaction. These switching techniques are now mainstream and are used in commercial products for non-volatile memory.

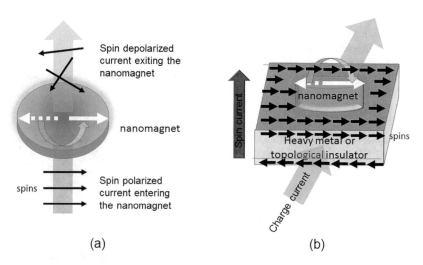

Fig. 3.1 Switching the magnetization of nanomagnets possessing in-plane anisotropy using **a** spin transfer torque, and **b** giant spin Hall effect in heavy metals or topological insulators

One important difference between STT on the one hand and GSHE-enhanced STT or SOT on the other hand, is that the former utilizes only two terminals (for current passage), while the latter requires at least three terminals (two for current passage and at least one additional one to generate GSHE or SOT). Even though the STT is less energy-efficient than either GSHE-embellished STT or SOT, it has found wider use in memory since the STT device has a smaller footprint (needing only two terminals instead of three). In memory, the high density consideration is primary and hence a 2-terminal device is always preferred over a 3-terminal device.

3.3   Voltage-Controlled Mechanisms for Switching the Magnetization of Nanomagnets

3.3.1   Voltage Controlled Magnetic Anisotropy (VCMA)

The VCMA mechanism refers to the change in the magnetic anisotropy of a nanomagnet upon the application of a voltage across its interface with another material. A large VCMA effect was observed first in Fe/MgO junctions [6, 7] and since then has been utilized to switch the resistance states of MTJs whose three layers are typically a synthetic antiferromagnet (hard layer), MgO (spacer layer) and Fe, Co, or CoFeB (soft layer). The ferromagnetic layers usually have perpendicular magnetic anisotropy[1] and hence their two stable magnetization orientations are perpendicular to their planes ("up" and "down"). When a voltage pulse is applied across the MTJ, the soft layer's magnetic anisotropy

---

[1] Such an MTJ is sometimes referred to as a p-MTJ because the ferromagnetic layers have perpendicular magnetic anisotropy.

Fig. 3.2 Voltage controlled magnetic anisotropy (VCMA) switching of a magnetic tunnel junction with perpendicular magnetic anisotropy (p-MTJ). This is precessional switching in the presence of an in-plane magnetic field

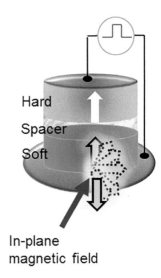

In-plane
magnetic field

changes from perpendicular to in-plane and that tends to rotate its magnetization from out-of-plane to in-plane as shown in Fig. 3.2. If there is an in-plane magnetic field, then the magnetization will not come to rest in-plane, but continue to rotate (or precess) about the in-plane magnetic field. If the voltage pulse width is the same as the time it takes to precess through 180°, then the magnetization precession will (ideally) stop when the soft layer's magnetization completes the 180° rotation, because then it has reached one of the two stable orientations for the magnetization. At that point the soft layer's magnetization has flipped and so has the resistance of the MTJ and the stored bit of information [8, 9].

There are two major inconveniences with VCMA switching. The first is the requirement for an in-plane magnetic field, and the second in the need to control the voltage pulse width precisely. If the pulse duration undershoots or overshoots the time required to precess through 180°, then the switching may fail. In any case, since this time is not fixed and can vary randomly in the presence of thermal noise, VCMA switching usually has a high error probability of the order of $\sim 10^{-5}$.

Various approaches have been taken to eliminate the externally applied in-plane magnetic field. One approach is to place a ferromagnetic layer on top of the MTJ that provides the in-plane magnetic field [10]. Another approach is to use a conically magnetized free layer in the shape of an elliptical disk [11]. There is a shape-anisotropy related effective magnetic field (sometimes called the "demagnetizing field") in the soft layer which can act as the in-plane magnetic field. Yet another approach is to build the p-MTJ on top of a piezoelectric layer and apply a voltage on the piezoelectric with a suitable placement of electrodes to generate uniaxial stress along the major or minor axis of the soft layer. The stress acts like an effective magnetic field that is in-plane [12].

Several mechanisms have been proposed to explain the origin of the VCMA effect, such as spin-dependent screening of the electric field at the ferromagnetic interface with the spacer layer [13], band structure modification in the soft layer due to mixing of p- and d-orbitals near the Fermi level [14], and the relative modification in the electron filling of

Table 3.1 Energy dissipated in different switching mechanisms

Switching mechanism	Energy dissipated
Magnetic field generated by an on-chip current	1–10 pJ
Spin transfer torque (STT)	10–100 fJ
Spin orbit torque (SOT)	1–10 fJ
Voltage controlled magnetic anisotropy (VCMA)	100 aJ–1 fJ
Field effect transistor	50 aJ–100 aJ
Straintronics	1–10 aJ

each 3d orbital causing charge accumulation in the magnetic layer [15]. Other proposed mechanisms are electric field-induced modification of the Rashba spin-orbit anisotropy [16, 17], atomic displacement at the soft layer interface with the spacer layer [18], and coupling of electric quadrupole with intra-atomic magnetic dipole [19]. A recent review of the VCMA mechanism can be found in ref. [20].

3.3.2 Straintronics

Straintronic switching is of interest because it is extremely energy-efficient. Table 3.1 provides a comparison of the energy dissipated in different switching mechanisms.

Based on this table, where the data have been collected from many reported experimental demonstrations and rigorous theoretical simulations, *the only mainstream magnet switching mechanism which seems to be more energy-efficient than switching a state-of-the-art field effect transistor is straintronics.* It therefore behooves us to study this mechanism minutely.

Magnetic straintronics refers to the science and technology of switching the magnetization of a *magnetostrictive* nanomagnet (e. g. Co, Ni, FeGa, Terfenol-D, etc.) using electrically generated mechanical strain by employing the inverse magnetostriction effect, also known as the Villari effect. A magnetostrictive single domain nanomagnet with in-plane anisotropy, whose magnetization has two stable states, is placed on top of a *poled* piezoelectric film grown on a conducting substrate which is grounded. The nanomagnet and the film are elastically coupled, forming a two phase (magnetostrictive+piezoelectric) multiferroic. The nanomagnet can be shaped like an elliptical disk or a rectangular slab so that it has two stable magnetization directions. Electrodes are placed around the nanomagnet in a suitable geometry. The spacing between the electrodes' edge and the nearest nenaomagnet edge, the lateral dimension of the nanomagnet and the electrodes, and the piezoelectric film thickness are all of the same order to ensure generation of biaxial strain in the piezoelectric region underneath the magnet with a spatial distribution favorable to switching the magnetization via the Villari effect [21, 22]. The configuration is shown in Fig. 3.3 for an elliptical nanomagnet.

When a voltage is applied to the electrodes (which are shorted together), a vertical electric field appears in the piezoelectric film and a biaxial strain is generated in the piezoelectric region pinched between the electrodes. This strained region is underneath the nanomagnet and is in elastic contact with it, so the strain is transferred partially or wholly to the nanomagnet. In Fig. 3.3, if the applied voltage generates an electric field whose direction is opposite to that of the poling, then compressive stress will be generated along the major axis (which the so-called "easy axis" of magnetization because the two bistable orientations lie along this direction) and tensile stress along the minor axis. If we reverse the voltage polarity (and hence the direction of the applied electric field), then the signs of the stresses will also be reversed.

Let σ be the uniaxial stress generated along the major axis and λ_s the magnetostriction coefficient of the nanomagnet material. The sign of the stress is positive if it is tensile

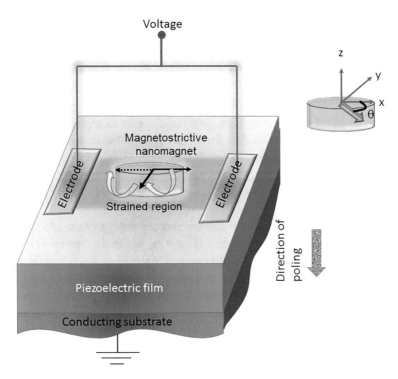

Fig. 3.3 Straintronic switching: Rotating the magnetization of a magnetostrictive elliptical nano-magnet, elastically coupled to an underlying poled piezoelectric film with a voltage applied across the piezoelectric. The nanomagnet has in-plane magnetic anisotropy. As long as the sign of the product of the uniaxial component of the stress along the major axis (easy axis) and the magnetostriction coefficient is negative, the stress will cause the magnetization's in-plane component to rotate away from the major axis towards the minor axis (hard axis)

and negative if it is compressive. As long as the product $\lambda_s\sigma$ is *negative*, the application of the stress, if it is of sufficient strength, will cause the magnetization of the nanomagnet to rotate away from the major axis (stable direction or easy axis) towards the minor axis (unstable direction or hard axis). If the stress is kept on, the magnetization will settle along the minor axis and remain there. Why this happens can be understood by examining the potential energy profile in the plane of the nanomagnet. The potential energy in the *plane* of the nanomagnet is given by [23]

$$E(\theta) = \left(\mu_0/2\right)M_s^2\Omega\left[N_{d-yy}\cos^2\theta + N_{d-xx}\sin^2\theta\right] - \left(3/2\right)\lambda_s\sigma\Omega\cos^2\theta \qquad (3.1)$$

where M_s is the saturation magnetization of the nanomagnet material, N_{d-xx} and N_{d-yy} are the demagnetization factors along the major and minor axes (which depend on the lateral dimensions and thickness of the nanomagnets [24]), Ω is the volume of the nanomagnet, μ_0 is the permeability of vacuum, and θ is the angle subtended by the magnetization with the major axis as shown in the inset of Fig. 3.3. The potential energy is plotted as a function of θ for three different values of the stress in Fig. 3.4. In the unstressed condition, or under low stress, the energy minima are at $\theta = 0°$, $180°$, implying that the stable orientation is along the major axis, but at high enough stress such that $|\sigma\lambda_s| > \mu_0 M_s^2\left[N_{d-yy} - N_{d-xx}\right]/3$, the energy minimum moves to $\theta = 90°$, at which point, the magnetization points along the minor axis or the hard axis.

When the stress is ultimately relaxed, the magnetization will return to the major axis, with *equal probability* of returning to the original orientation and the opposite orientation. This is disappointing since switching can then be accomplished with only 50% probability, which is, of course, unacceptable.

This problem seems to go away if one could encode the binary bit 0 in the orientation along the major axis and the binary bit 1 in the orientation along the minor axis, or vice versa. In that case, we can switch with much higher than 50% probability (ideally 100% probability), since sufficiently high stress will almost always move the magnetization to the hard axis and relaxing the stress will move it back to the easy axis. Unfortunately, in that case, the stress has to be *kept on* to retain the magnetization pointing along the

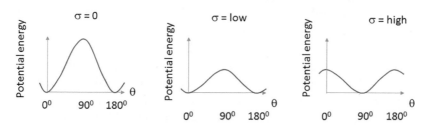

Fig. 3.4 Potential energy profile in the plane of the elliptical magnetostrictive nanomagnet for three different values of the uniaxial stress generated along the major axis. The product $\lambda_s\sigma$ is assumed to be negative

minor axis. If the voltage generating the stress is withdrawn, the stress will relax and the magnetization will *not* remain pointing along the minor axis but return to the major axis, which means that such a switch is *volatile* and does not retain information. There is, of course, some possibility that the stress may *not* disappear when the voltage generating it is withdrawn (non-volatile stress). In this case, a voltage of opposite polarity will be required to make it disappear. That indeed would implement a non-volatile switch but it will be rather unreliable since the non-volatile stress in the piezoelectric could relax spontaneously under temperature fluctuations and other extraneous influences.

To make the switch *reliably non-volatile*, we must encode the binary bits in the two stable orientations along the major axis (easy axis), and then find a way of rotating the magnetization by 180°, not just 90°. Various methodologies to accomplish that are discussed in the next chapter.

3.4 Historical Perspective

This chapter is concluded by providing a historical perspective on switching the magnetization of a magnetostrictive nanomagnet with electrically generated mechanical strain. The dawn of magnetic straintronics goes back to at least 1995 [25]. An attempt was made to switch the magnetization of a thin nickel film (a magnetostrictive metal) deposited on a weakly piezoelectric GaAs substrate, but it did not succeed owing to the weak piezoelectricity of GaAs. This was the beginning of straintronic switching with two-phase multiferroics.

The use of piezoelectric materials with stronger piezoelectricity soon produced numerous reports of controlling the magnetization in magnetostrictive films deposited on piezoelectric films using voltage-generated strain [26]. These experiments showed reversible control of nanomagnetic domains [27], repeatable reversal of perpendicular magnetization in the absence of a magnetic field in regions of a Ni film [28], and reversal of perpendicular magnetization in Co/Ni multilayers [29]—all caused by strain. Strain control of magnetization orientation in manganite titanate [30] and lanthanum strontium manganite (LSMO) films [31], iron films [32], $TbCo_2$/FeCo multilayers [33] and strain control of magnetic properties of FeGa/NiFe multilayer films [34] as well as FeGa films [35] have also been reported. Strain has been shown to reorient magnetization in Ni rings [36, 37] and Ni squares of 2 microns side [38] and the soft layer of MTJs of lateral dimensions 20 μm $\times$ 40 μm [39]. This latter effect has been utilized to read the magnetization orientation in a composite multiferroic heterostructure $[Ni(TbCo_2/FeCo)]/[Pb(Mg_{1/3}Nb_{2/3})O_3]_{1-x}[PbTiO_3]_x$ [40].

There are many reports of strain mediated manipulation of the magnetization in nanomagnets deposited on piezoelectric substrates. For example, an electric field induced stress mediated reversible control of magnetization orientation in nanomagnets of nominal lateral dimensions 380 nm $\times$ 150 nm deposited on a 1.28 micron PZT thin film was demonstrated with the application of a voltage of 1.5 V to the PZT film [40]. Control of magnetoelectric heterostructures with localized strain to reorient the magnetization in a Ni ring of 1000 nm outer diameter, 700 nm inner diameter, and 15 nm thickness, ultimately lead to the deterministic multistep reorientation of magnetization in a 400 nm Ni dot of 15 nm thickness [41].

Uniform magnetization rotation through 90° has also been demonstrated through imaging with X-ray photoemission electron microscopy (X-PEEM) and X-ray magnetic circular dichroism (XMCD) in elliptical nanomagnets of nominal lateral dimensions ~100 nm × 150 nm [42]. Thus, the basic notion behind straintronics has been around for almost three decades but only recently it entered the limelight after its promise of exceptional energy-efficiency was appreciated and understood.

References

1. M.T. Alam, M.J. Siddiq, G.H. Bernstein, M. Niemier, W. Porod, X.S. Hu, On-chip clocking for nanomagnetic logic devices. IEEE Trans. Nanotech. **9**, 348 (2010)
2. See, for example, D.C. Ralph, M.D. Stiles, Spin transfer torques. J. Magn. Magn. Mater. **320**, 1190 (2008) and references therein
3. See, for example, L. Liu, C.-F. Pai, H.W. Tseng, D.C. Ralph, R.A. Buhrman, Spin torque switching with giant spin Hall effect. Science **336**, 555 (2012)
4. See, for example, A.R. Mellnik, J.S. Lee, A. Richardella, J.L. Grab, P.J. Mintun, M.H. Fischer, A. Vaezi, A. Manchon, E.A. Kim, N. Samarth, D.C. Ralph, Spin transfer torque generated by a topological insulator. Nature **511**, 449 (2014)
5. I.M. Miron, K. Garello, G. Gaudin, P.-J. Zermatten, M.V. Costache, S. Auffret, S. Bandiera, B. Rodmacq, A. Schuhl, P. Gambardella, Perpendicular switching of a single ferromagnetic layer induced by in-plane current injection. Nature **476**, 189 (2011)
6. T. Maruyama, Y. Shiota, T. Nozaki, K. Ohta, N. Toda, M. Mizuguchi, A.A. Tulapurkar, T. Shinjo, M. Shiraishi, S. Mizukami, Y. Ando, Y. Suzuki, Large voltage-induced magnetic anisotropy change in a few atomic layers of iron. Nat. Nanotech. **4**, 158 (2009)
7. T. Nozaki, Y. Shiota, M. Shiraishi, T. Shinjo, Y. Suzuki, Voltage induced perpendicular magnetic anisotropy change in magnetic tunnel junctions. Appl. Phys. Lett. **96**, 022506 (2010)
8. S. Kanai, M. Yamanouchi, S. Ikeda, Y. Nakatani, F. Matsukura, H. Ohno, Electric field induced magnetization reversal in a perpendicular-anisotropy CoFeB-MgO magnetic tunnel junction. Appl. Phys. Lett. **101**, 122403 (2012)
9. Y. Shiota, T. Nozaki, F. Bonell, S. Murakami, T. Shinjo, Y. Suzuki, Induction of coherent magnetization switching in few atomic layers of FeCo using voltage pulses. Nat. Mater. **11**, 39 (2012)
10. Y. Wu, W. Kim, K. Grarello, F. Yasin, G. Jayakumar, S. Couet, R. Carpenter, S. Kundu, S. Rao, D. Crotti, J. V. Houdt, G. Groeseneken, G. Kar, Deterministic and field-free voltage-controlled MRAM for high performance and low power applications, in *2020 IEEE Symposium on VLSI Technology* (2020), p. 20237270
11. R. Matsumoto, T. Nozaki, S. Yuasa, H. Imamura, Voltage-induced precessional switching at zero-bias magnetic field in a conically magnetized free layer. Phys. Rev. Appl. **9**, 014026 (2018)

12. J.L. Drobicth, M.A. Abeed, S. Bandyopadhyay, Precessional switching of a perpendicular anisotropy magneto-tunneling junction without a magnetic field. Jpn. J. Appl. Phys. (Rapid Commun.) **56**, 100309 (2017)
13. C.G. Duan, J.P. Velev, R.F. Sabirianov, Z. Zhu, J. Chu, S.S. Jaswal, E.Y. Tsymbal, Surface magnetoelectric effect in ferromagnetic metal films. Phys. Rev. Lett. **101**, 137201 (2008)
14. K. Nakamura, R. Shimabukuro, Y. Fujiwara, T. Akiyama, T. Ito, A.J. Freeman, Giant modification of the magnetocrystalline anisotropy in transition-metal monolayers by an external electric field. Phys. Rev. Lett. **102**, 187201 (2009)
15. M. Tsujikawa, T. Oda, Finite electric field effects in the large perpendicular magnetic anisotropy surface Pt/Fe/Pt(001): a first-principles study. Phys. Rev. Lett. **102**, 247203 (2009)
16. L. Xu, S. Zhang, Electric field control of interface magnetic anisotropy. J. Appl. Phys. **111**, 07C501 (2012)
17. S. Barns, J. Ieda, S. Maekawa, Rashba spin-orbit anisotropy and the electric field control of magnetism. Sci. Rep. **4**, 4105 (2014)
18. K. Nakamura, T. Akiyama, T. Ito, M. Weinert, A. Freeman, Role of an interfacial FeO layer in the electric-field-driven switching of magnetocrystalline anisotropy at the Fe/MgO interface. Phys. Rev. B **81**, 220409(R) (2010)
19. S. Miwa, M. Suzuki, M. Tsujikawa, K. Matsuda, T. Nozaki, K. Tanaka, T. Tsukahara, K. Nawaoka, M. Goto, Y. Kotani, T. Ohkubo, F. Bonell, E. Tamura, K. Hono, T. Nakamura, M. Shirai, S. Yuasa, Y. Suzuki, Voltage controlled interfacial magnetism through platinum orbits. Nat. Commun. **8**, 15848 (2017)
20. T. Yamamoto, R. Matsumoto, T. Nozaki, H. Imamura, S. Yuasa, Developments in voltage-controlled sub-nanosecond magnetization switching. J. Magn. Magn. Mater. **560**, 169637 (2022)
21. J. Cui, J.L. Hockel, P.K. Nordeen, D.M. Pisani, C.-Y Liang, G.P. Carman, C.S. Lynch, Appl. Phys. Lett. **103**, 232905 (2013)
22. C.Y. Liang, S.M. Keller, A.E. Sepulveda, W.Y. Sun, J.Z. Cui, C.S. Lynch, G.P. Carman, J. Appl. Phys. **116**, 123909 (2014)
23. K. Roy, S. Bandyopadhyay, J. Atulasimha, Energy dissipation and switching delay in stress-induced switching of multiferroic nanomagnets in the presence of thermal fluctuations. J. Appl. Phys. **112**, 023914 (2012)
24. S. Chikazumi, *Physics of Magnetism* (Wiley, New York, 1964)
25. R.I. Dzhioev, B.P. Zakharchenya, V.L. Korenev, Optical orientation study of thin ferromagnetic films in a ferromagnetic/semiconductor structure. Fiz. Tverd. Tela **37**, 3510 (1995)
26. W. Eerenstein, N.D. Mathur, J.F. Scott, Multiferroic and magnetoelectric materials. Nature **442**, 759 (2016)
27. T. Brintlinger, S. Lim, K. H. Baloch, P. Alexander, Y. Qi, J. Barry, J. Melngailis, L. Salamanca-Riba, I. Takeuchi, J. Cummings, In-situ observation of reversible nanomagnetic switching induced by electric fields. Nano Lett. **10**(4), 1219 (2010)
28. M. Ghidini, R. Pellicelli, J.L. Prieto, X. Moya, J. Soussi, J. Briscoe, S. Dunn, N.D. Mathur, Non-volatile electrically driven repeatable magnetization reversal with no applied magnetic field. Nat. Commun. **4**, 1453 (2013)
29. D.B. Gopman, C.L. Dennis, P.J. Chen, Y.L. Iunin, P. Finkel, M. Staruch, R.D. Shull, Strain assisted magnetization reversal in Co/Ni multilayers with perpendicular magnetic anisotropy. Sci. Rep. **6**, 27774 (2016)
30. R.V. Chopdekar, J. Heidler, C. Piamonteze, Y. Takamura, A. Scholl, S. Rusponi, H. Brune, L.J. Heyderman, F. Nolting, Strain-dependent magnetic configurations in manganite-titanate heterostructures probed with soft X-ray techniques. Eur. Phys. J. B **86**, 241 (2013)

31. J. Heidler, C. Piamonteze, R.V. Chopdekar, M.A. Uribe-Laverde, A. Alberca, M. Buzz, A. Uldry, B. Delley, C. Bernhard, F. Nolting, Manipulating magnetism in $La_{0.7}Sr_{0.3}MnO_3$ via piezostrain. Phys. Rev. B **91** 024406 (2015)
32. G. Venkataiah, Y. Shirahata, I. Suzuki, M. Itoh, T. Taniyama, Strain-induced reversible and irreversible magnetization switching in $Fe/BaTiO_3$ heterostructures. J. Appl. Phys. **111**, 033921 (2012)
33. Y. Dusch, N. Tiercelin, A. Klimov, S. Giordano, V. Preobrazhensky, P. Pernod, Stress-mediated magnetoelectric memory effect with uni-axial $TbCo_2/FeCo$ multilayer on 011-cut PMN-PT ferroelectric relaxor. J. Appl. Phys. **113**, 17C719 (2013)
34. C.R. Rementer, K. Fitzell, Q. Xu, P. Nordeen, G.P. Carman, Y.E. Wang, J.P. Chang, Tuning static and dynamic properties of FeGa/NiFe heterostructures. Appl. Phys. Lett. **110**, 242403 (2017)
35. G. Dai, Q. Zhan, Y. Liu, H. Yang, X. Zhang, B. Chen, R.-W. Li, Mechanically tunable magnetic properties of $Fe_{81}Ga_{19}$ films grown on flexible substrates. Appl. Phys. Lett. **100**, 122407 (2012)
36. E.C. Corredor, D. Coffey, M. Ciria, J.I. Arnaudas, J. Aisa, C.A. Ross, Strain-induced magnetization reorientation in epitaxial Cu/Ni/Cu rings. Phys. Rev. B **88**, 054418 (2013)
37. H. Sohn, M.E. Nowakowski, C.-Y. Liang, J.L. Hockel, K. Wetzlar, S. Keller, B.M. McLellan, M.A. Marcus, A. Doran, A. Young, M. Kläui, G.P. Carman, J. Bokor, R.N. Candler, Electrically driven magnetic domain wall rotation in multiferroic heterostructures to manipulate suspended on-chip magnetic particles. ACS Nano **9**, 4814 (2015)
38. S. Finizio, M. Foerster, M. Buzzi, B. Krüger, M. Jourdan, C.A.F. Vaz, J. Hockel, T. Miyawaki, A. Tkach, S. Valencia, F. Kronast, G.P. Carman, F. Nolting, M. Kläui, Magnetic anisotropy engineering in thin film Ni nanostructures by magnetoelastic coupling. Phys. Rev. Appl. **1**, 1 (2014)
39. A. Klimov, N. Tiercelin, Y. Dusch, S. Giordano, T. Mathurin, P. Pernod, V. Preobrazhensky, A. Churbanov, S. Nikitov, Magnetoelectric write and read operations in a stress-mediated multiferroic memory cell. Appl. Phys. Lett. **110**, 222401 (2017)
40. T.K. Chung, S. Keller, G.P. Carman, Electric-field-induced reversible magnetic single-domain evolution in a magnetoelectric thin film. Appl. Phys. Lett. **94**, 132501 (2009)
41. J. Cui, C.-Y. Liang, E.A. Paisley, A. Sepulveda, J.F. Ihlefeld, G.P. Carman, C.S. Lynch, Generation of localized strain in a thin film piezoelectric to control individual magnetoelectric heterostructures. Appl. Phys. Lett. **107**, 092903 (2015)
42. H. Sohn, C.-Y. Liang, M.E. Nowakowski, Y. Hwang, S. Han, J. Bokor, G.P. Carman, R.N. Candler, Deterministic multi-step rotation of magnetic single-domain state in Nickel nanodisks using multiferroic magnetoelastic coupling. J. Magn. Magn. Mater **439**, 196 (2017)

Full Magnetic Reversal of a Magnetostrictive Nanomagnet Using Electrically Generated Strain

4

As discussed in the previous chapter, strain will usually rotate the magnetization of a magnetostrictive nanomagnet by up to $90°$, which is not the most desirable outcome. Rotation by full $180°$ is preferred for two reasons. First, as explained in the previous chapter, it makes the switch non-volatile. Second, the magnetization state will usually be read with a magnetic tunnel junction (MTJ) via its resistance, and the resistance difference between the two magnetization orientations is much larger if the angular separation between them is $180°$ rather than $90°$.

Let us call the MTJ resistance in the parallel state (i.e., when the magnetizations of the hard and soft layers are mutually parallel) R_P and that in the antiparallel state R_{AP}. The resistance, when the angle between the magnetizations of the two layers is θ, is given by $R_{MTJ} = R_P + \frac{R_{AP}-R_P}{2}[1 - \cos\theta]$, which is $\frac{R_{AP}+R_P}{2}$ when $\theta = 90°$. The resistance ratio, when the angular separation is $180°$, is $\rho_{180°} = R_{AP}/R_P$, whereas when it is $90°$, the resistance ratio is $\rho_{90°} = R_{AP}/2R_P + 1/2$. Clearly $\rho_{180°}$ is always larger than $\rho_{90°}$ since $R_{AP} > R_P$. This is why an angular separation of $180°$ is preferred.

Fortunately, there are many ways of making the magnetization rotate by full $180°$ using electrically generated strain. Some of them are discussed below along with their pros and cons.

4.1 Precisely Controlled Strain Pulse for Complete Magnetization Reversal

One simple way to generate $180°$ rotation (i.e. flip the magnetization from one stable orientation along the major axis of an elliptical nanomagnet to the other stable orientation in the opposite direction) is to apply a uniaxial strain of the appropriate sign (the product $\lambda_s\sigma$ has to be *negative*) along the major axis. The nanomagnet is assumed to have in-plane magnetic anisotropy. The stress will begin to rotate the magnetization's in-plane component towards the minor axis, but then we must remove the strain *as soon as*

© The Author(s), under exclusive license to Springer Nature Switzerland AG 2022
S. Bandyopadhyay, *Magnetic Straintronics*, Synthesis Lectures on Engineering, Science, and Technology, https://doi.org/10.1007/978-3-031-20683-2_4

the in-plane component has rotated through 90° and aligned with the minor axis. During the rotation, the magnetization vector actually lifts somewhat out of the plane of the nanomagnet because that reduces the stress anisotropy energy. However, the out-of-plane excursion also increases the shape anisotropy energy and the hence a demagnetizing field is generated which tries to bring the magnetization back to the plane to reduce the shape anisotropy energy. This field is strongest when the magnetization enters the plane perpendicular to the major axis since there the stress anisotropy energy vanishes and only the shape anisotropy energy remains. The demagnetizing field creates a residual torque on the magnetization. Depending on the location of the magnetization when it enters the plane perpendicular to the major axis, the residual torque will either continue to rotate the magnetization toward the desired orientation opposite to that of the initial orientation (switching success) or the opposite direction (switching failure) [1]. Ref. [1] showed that because of the complex interactions between the shape anisotropy and stress anisotropy, both outcomes are not equally likely. The residual torque is favorable more often than not and hence the probability of switching successfully is more than 50%. The disadvantage of this approach is obvious; one cannot remove the strain prematurely before completing 90° rotation of the magnetization's in-plane component, nor can one procrastinate and not remove the strain immediately after completing the 90° rotation. *The strain must be withdrawn at precisely the right juncture.* If one lingers with the strain after completing the 90° rotation, the magnetization will relax into the plane of the nanomagnet to minimize the shape anisotropy energy, thereby annihilating the demagnetizing field and the residual torque responsible for rotating the magnetization beyond 90°. In that case, the magnetization will actually settle along the minor axis of the ellipse (the hard axis) and upon removal of stress will return to one of the two stable orientations along the major axis with equal likelihood. The switching probability in this scenario will be 50%, which is far from deterministic switching. Therefore, *precise timing of the pulse* is needed to accomplish the full magnetization reversal with high probability of success well exceeding 50%. This is reminiscent of the case of precessional switching via the voltage controlled magnetic anisotropy (VCMA) approach.

There are two additional problems with the above approach. First, the strain cannot be removed instantaneously since piezo response has a finite time duration and hence the ramp down rate of the stress matters. Obviously, the ramp down rate must be high for the error probability to be relatively low. Second, the time to complete the 90° rotation is not a fixed quantity at room temperature when thermal noise is present. There is some variability in this time, which would make it impossible to withdraw the stress at the right moment always. Thus, thermal noise will increase the switching error rate in this modality of straintronic switching, as in the case of VCMA. However, with careful design, high ramp down rate for the stress, and high enough stress value, the switching error probability at room temperature can be made reasonably low ($\sim 10^{-5}$), as was shown in [1].

The energy dissipated to flip the magnetization of a magnetostrictive nanomagnet with strain was computed in the presence of thermal noise at room temperature using the stochastic Landau-Lifshitz-Gilbert (sometimes referred to as the Landau-Lifshitz-Gilbert-Langevin) equation in Ref. [2]. The energy dissipation has two components: the

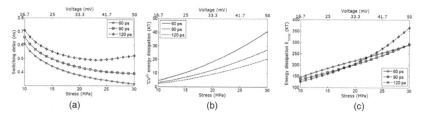

Fig. 4.1 **a** Switching delay, **b** the extrinsic CV^2 energy dissipation, and **c** total (intrinsic + extrinsic) energy dissipation as functions of stress (plotted along lower horizontal axis) and the voltage needed to generate that stress (plotted along the upper horizontal axis) for three different (linear) ramp durations for the magnetostrictive nanomagnet/piezoelectric combination (two-phase multiferroic) discussed in the text. Note that the nanomagnet can be switched in less than 1 ns while dissipating energy as low as 125 kT at room temperature, which is roughly 0.5 aJ, which is two orders of magnitude smaller than the energy dissipated to switch a state-of-the-art nano-transistor at the time of writing. Reproduced from *J. Appl. Phys.*, **112**, 023,914 (2012) with the permission of AIP Publishing

intrinsic dissipation in the nanomagnet due to Gilbert damping and the extrinsic dissipation associated with the external circuit which generates and applies the voltage on the piezoelectric with the electrodes shown in Fig. 3.3. This latter component is roughly equal to $(1/2)CV^2$ where C is the capacitance of the electrodes and V is the voltage applied to the electrodes to generate the stress required to switch. Figure 4.1 shows the dissipation plots. These plots were generated assuming that the nanomagnet is made of Terfenol D with major axis = 100 nm, minor axis = 90 nm and thickness = 6 nm. These dimensions make the internal (shape-anisotropy) energy barrier within the nanomagnet, accruing from its elliptical shape, 44 kT at room temperature [2]. The piezoelectric layer was assumed to be lead zirconate titanate (PZT) whose thickness was assumed to be four times the nanomagnet thickness. The piezoelectric parameters for bulk PZT were used, although, in reality, these parameters are likely to be significantly degraded when the PZT layer thickness is a mere 24 nm. The energy dissipations were plotted for various rates of ramping up the voltage and the ramp down rate was assumed to be the same as the ramp-up rate.

4.2 Successive 90° Rotations for Complete Magnetization Reversal

Alternate approaches for 180° rotation have been proposed by others. Reference [3] theoretically analyzed a scenario where uniaxial stress of the appropriate sign (to make $\lambda_s \sigma$ negative) is applied at an angle to the easy axis in a nanomagnet with four fold symmetry (shaped like a leaf) which rotates the magnetization by an angle ϕ. Subsequently, the stress is changed in sign which rotates the magnetization further by an additional $90°$. The stress is then relaxed and the magnetization rotates to the nearest stable orientation which is antiparallel to the initial orientation, thereby completing the 180° rotation. This

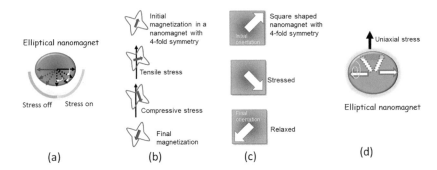

Fig. 4.2 Different modalities of switching the magnetization of a magnetostrictive nanomagnet through 180° using electrically generated strain using the inverse magnetostriction (Villari) effect: **a** Switching the magnetization of an elliptical nanomagnet with two stable magnetization orientations using a precisely timed stress pulse, **b** switching the magnetization of a leaf-shaped nanomagnet with four stable magnetization orientations using successive 90° rotation and two different signs of the stress, **c** switching the magnetization of a square-shaped nanomagnet with four stable magnetization orientations using successive 90° rotation and a single stress (of one sign), and **d** switching the magnetization of an elliptical nanomagnet with two stable magnetization orientations using a precisely time stress pulse and precessional dynamics

is shown in Fig. 4.2b. A simpler version was proposed in Ref. [4] where anisotropic strain applied to a square shape nanomagnet (again with four-fold symmetry or four stable orientations) rotates the magnetization by 90° and then another 90° rotation takes place after removal of stress, resulting in full 180^0 rotation. The principal component of the strain generated in this case is roughly along a diagonal of the square and of course the $\lambda_s \sigma$ product is negative. These mechanisms do not require precisely timed stress pulses, which is an advantage, but they require structures with fourfold symmetry whose magnetization states are not bistable.

Reference [5] proposed yet another approach. It suggested applying a uniaxial strain pulse of the appropriate sign along the minor axis (hard axis) of an elliptical nanomagnet which will induce the magnetization to rotate from its initial orientation along the major axis (easy axis) to the hard axis and thereafter precess around it as long as the damping is low enough. If the strain is turned off while the precessing magnetization has a component along the easy axis, but in the direction opposite to that of the initial orientation, the magnetization will move towards that orientation, precess around it and ultimately settle into that orientation as shown in Fig. 4.2, resulting in 180° switching. Like the strategy in Ref. [1], this also requires precise timing of the stress pulse, but in this case, the nanomagnet can be shaped like an elliptical disk and have only two stable orientations. In essence, this method is an exact analog of that in Ref. [1] and has the same advantages and disadvantages.

The successive rotation mechanisms described in this subsection have been studied with theoretical simulations and at the time of this writing, no experimental demonstration has been reported. Since these simulations were carried out for 0 K temperature (no account was made for thermal noise), the switching error probability at room temperature is not known. The strategy is Ref. [5], however, is analogous to that in Ref. [1] and therefore is likely to have similar error probability at room temperature, namely ~10^{-5}.

4.3   Switching with a 4-Electrode Configuration for Complete Magnetization Reversal

Another strategy to effect complete 180° switching with strain is to apply two uniaxial stresses on an elliptical nanomagnet sequentially in two different directions (neither of which is along the easy or hard axis) and that can complete the 180° rotation in two steps [6]. For this purpose, two electrode pairs are deposited on a poled piezoelectric film surrounding a magnetostrictive nanomagnet in a geometry where the two lines joining opposite electrodes subtend an acute angle with each other. This is shown in the top panel of Fig. 4.4. Each of the two members in a pair are electrically shorted and a common voltage is applied to them.

One pair is first activated by applying a voltage of a certain polarity to it which generates biaxial stress in the piezoelectric, say, compressive along the line joining the two members of the pair and tensile in the perpendicular direction. One can roughly approximate the effect of this biaxial stress by assuming that uniaxial compressive stress is generated along the line joining the two activated electrodes. Such a stress will rotate the magnetization away from the major axis (easy axis) of the nanomagnet and roughly stabilize it in a direction perpendicular to the line joining the activated pair if the magnetostriction of the nanomagnet is positive so that the $\lambda_s\sigma$ product has a negative sign. If the magnetostriction is negative, then one simply needs to reverse the voltage polarity to generate stresses of the opposite sign.

After activating the first pair which produces a rotation of the magnetization through an *acute* angle ϕ, the second pair is activated, followed by deactivation of the first pair, and this rotates the magnetization through an additional angle, bringing the total rotation to an *obtuse* angle θ, where $90° < \theta < 180°$. Finally when the second electrode pair is deactivated, the magnetization relaxes to the *nearest* stable orientation, which is along the major axis but pointing opposite to the initial direction. That completes $180°$ rotation, or full magnetization reversal. This strategy would require four electrode pads as shown in the top panel of Fig. 4.4 and therefore produces a relatively large footprint. However, the advantage is that one does not need any precisely timed pulse and the magnet also has two fold symmetry and is hence bistable, instead of multi-stable.

This switching mechanism has been demonstrated experimentally [7] (see bottom panel of Fig. 4.3) using Co nanomagnets delineated on a piezoelectric (001) PMN-PT substrate.

What immediately stands out from the experiment is that the switching error probability is extremely large since no more than one out of four nanomagnets actually switched, which means that the error probability is at least 75%. Theoretically, the error probability was estimated to be as low as 10^{-6} at room temperature, whereas the actual probability observed in experiments is several orders of magnitude larger! The reason for this is that the theoretical simulations

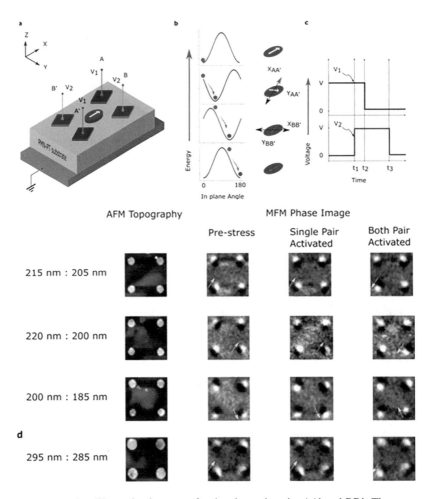

Fig. 4.3 (Top panel) **a** Electrode placement for the electrode pairs AA' and BB'. The nanomagnet is placed surrounded by the electrodes. **B** Potential energy profile of the nanomagnet as a function of the angle subtended by the magnetization with the major axis (direction of the arrow shown)—with no electrode activated, only AA' activated, BB' activated and AA' de-activated, and all electrodes de-activated. **c** Timing sequence of the voltage V_1 which is applied to electrode pair AA' and V_2 which is applied to electrode pair BB'. (Bottom panel). Atomic and magnetic force micrographs of four different sets of Co nanomagnet assemblies (with different major and minor axes dimensions) on a PMN-PT substrate subjected to this stress sequence. d. The magnetic force micrographs are shown at three different stages of electrode activation. One out of four nanomagnets in the set examined flips completely ($180°$ rotation) upon completion of the stress cycle, showing that the switching is error-prone (no more than 25% success probability). The flipped nanomagnet is designated with a yellow arrow. Reproduced with permission from Nano Letters, **17**, 3478 (2017) with permission of the American Chemical Society

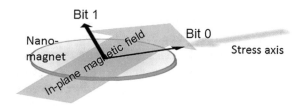

Fig. 4.4 Straintronic non-toggle memory

accounted only for errors caused by thermal noise. A much more lethal source of error is structural imperfections such as edge roughness and surface roughness and these can cause error probabilities of several percent. The high error probability is the most serious drawback of straintronic switching and detracts from its attractiveness as an extremely energy-efficient paradigm. Unfortunately, there always seems to be an inevitable trade-off between energy efficiency and reliability, and straintronic switching cannot avoid it either. We will discuss more of this in a later chapter.

4.4 Non-toggle Switch

The straintronic switches discussed so far are all *toggle switches*. Strain's only role is to flip the bistable magnetization from one stable orientation to the other and thereby "toggle" the switch. This is not particularly useful for writing bits in a memory cell (made of a nanomagnet) deterministically. Suppose we decide to write the bit 1 by orienting the magnetization in a particular (stable) direction. What we will do is: first read the stored bit (the existing magnetization direction) with an MTJ to see if it is already bit 1 or not. If it is already bit 1, we will do nothing. If not, we will flip the bit to change it from 0 to 1. That is all that a toggle switch can do. The point to note is that the "write" operation must be always preceded by a "read" operation in order to write the desired bit deterministically.

This is inconvenient to say the least. It would be nice to be able to write the bit deterministically without having to read it first. Such a switch is called a "non-toggle switch". The well-known example of this is the transistor where one polarity of the gate voltage always turns it on and the opposite polarity turns it off, which allows the writing of either bit deterministically, regardless of what the preceding bit state was. There is a non-toggle version of a straintronic switch as well [8] but it has two undesirable features. First, it needs an in-plane magnetic field to work, and second, the two magnetization orientations encoding bits 0 and 1 are not mutually antiparallel. Instead, they are mutually perpendicular, and as already discussed earlier, this reduces the distinguishability between the bits (i.e. reduces the resistance on/off ratio) when an MTJ is used to read them via spin-to-charge conversion.

The non-toggle straintronic switch consists of an elliptical nanomagnet (with in-plane magnetic anisotropy) where an in-plane magnetic field of the right strength is applied along the minor axis (hard axis) to bring the two stable orientations out of the major axis and make them point in two in-plane directions which are mutually perpendicular, as shown in Fig. 4.4. These two orientations encode the bits 0 and 1.

Uniaxial stress is then applied along one of the stable directions. If the sign of the stress (tensile is positive and compressive is negative) is such that the product of magnetostriction and stress, i.e., the $\lambda_s \sigma$ product is *negative*, then the magnetization will point along the other stable direction which has an angular separation of 90° from the stress axis. If, on the other hand, the $\lambda_s \sigma$ product is *positive*, the magnetization will point along the stress axis. Hence, by simply choosing the sign of the stress (compressive or tensile), one can write either bit 0 or bit 1 deterministically, *without having to read the stored bit first*. That makes it a "non-toggle switch". This strategy also has the advantage that no precise timing of the stress cycle is needed. Hence, we will expect that the switching error probability, or the write error probability (WEP), will be lower than the mechanisms that require precise timing. Unfortunately, this is not necessarily true since there are some other sources of switching error that are peculiar to this paradigm. If the in-plane magnetic field is misaligned with the minor axis, then the error rate climbs. Precise alignment is always challenging. The alignment may also vary across a wafer, causing randomness.

Finally, if the magnetization direction (or, equivalently, the bit state in this non-toggle switch) is "read" with a magnetic tunnel junction (MTJ), then the distinguishability between the bits will be relatively poor. This is because the resistance on/off ratio of the MTJ will be low owing to the fact that the two magnetization orientations corresponding to the "on" and "off" states are not mutually antiparallel. Instead, they are mutually perpendicular. Later it was shown that this can be ameliorated somewhat by using two pairs of electrodes instead of one to apply the stress [9] and that can allow the angular separation between the two stable directions to exceed 90°, resulting in a somewhat larger on/off ratio.

References

1. K. Roy, S. Bandyopadhyay, J. Atulasimha, Binary switching in a symmetric potential landscape. Sci. Rep. **3**, 3038 (2013)
2. K. Roy, S. Bandyopadhyay, J. Atulasimha, Energy dissipation and switching delay in stress-induced switching of multiferroic nanomagnets in the presence of thermal fluctuations. J. Appl. Phys. **112**, 023914 (2012)
3. J.J. Wang, J.M. Hu, J. Ma, J.X. Zhang, L.Q. Chen, C.W. Nan, Full 180^0 magnetization reversal with electric fields. Sci. Rep. **4**, 7507 (2014)
4. R.C. Peng, J.J. Wang, J.-M. Hu, L.-Q. Chen, C.-W. Nan, Electric field driven magnetization reversal in square shaped nanomagnet-based multiferroic heterostructure. Appl. Phys. Lett. **106**, 142901 (2015)
5. R.C. Peng, J.-M. Hu, K. Momeni, J.-J. Wang, L.-Q. Chen, C.W. Nan, Fast 180 magnetization switching in a strain-mediated multiferroic heterostructure driven by a voltage. Sci. Rep. **6**, 27561 (2016)

6. A.K. Biswas, S. Bandyopadhyay, J. Atulasimha, Complete magnetization reversal in a magnetostrictive nanomagnet with voltage-generated stress: a reliable energy-efficient non-volatile magneto-elastic memory. Appl. Phys. Lett. **105**, 072408 (2014)
7. A.K. Biswas, H. Ahmad, J. Atulasimha, S. Bandyopadhyay, Experimental demonstration of complete 180 reversal of magnetization in isolated Co nanomagnets on a PMN-PT substrate with voltage generated strain. Nano Lett. **17**, 3478 (2017)
8. N. Tiercelin, Y. Dusch, V. Preobrazhensky, P. Pernod, Magnetoelectric memory using orthogonal magnetization states and magnetoelastic switching. J. Appl. Phys. **109**, 07D726 (2011)
9. A.K. Biswas, S. Bandyopadhyay, J. Atulasimha, Energy-efficient magnetoelastic non-volatile memory. Appl. Phys. Lett. **104**, 232403 (2014)

Non-volatile Memory Implemented with Straintronic Magnetic Tunnel Junctions

5

It was mentioned earlier that the basic magnetic element that allows reading the stored bit (magnetization orientation) in a nanomagnet *by electrical means* is the magnetic tunnel junction (MTJ). The bit is actually stored in the magnetization orientation of the MTJ's soft layer and the MTJ resistance (high or low) tells us what that orientation (or stored bit) is. If that orientation is changed with electrically generated strain, we will call that MTJ a "straintronic MTJ". Note that the straintronic MTJ will have at least three terminals (usually four as shown in Fig. 5.1) and hence has a larger footprint than the basic MTJ which is two-terminal.

In a straintronic MTJ, the spin-to-charge converter consisting of the two ferromagnetic layers and the spacer, is delineated on a poled piezoelectric substrate (or a poled piezoelectric film deposited on a conducting substrate) with the soft layer in contact with the piezoelectric. Strain is generated in the soft layer by applying a voltage to electrodes flanking the MTJ on the piezoelectric surface. The metallic under layer and the metallic cap in Fig. 5.1 are for making contacts and also to prevent oxidation of the ferromagnetic layers. The metal under layer is also used for adhesion of the soft layer to the piezoelectric layer. It is thin enough to not seriously impede strain transfer from the piezoelectric to the soft layer.

5.1   Volatile Memory with Straintronic MTJ

Although there are methods to completely flip the magnetization of the soft layer with strain (180° rotation), as discussed in the previous chapter, most experimental demonstrations of straintronic MTJs have employed no more than 90° rotation since it is much simpler to achieve. Such MTJs can obviously implement only *volatile* memory cells since a 90° rotation will align the magnetization along the hard axis, and when the stress relaxes,

© The Author(s), under exclusive license to Springer Nature Switzerland AG 2022
S. Bandyopadhyay, *Magnetic Straintronics*, Synthesis Lectures on Engineering, Science, and Technology, https://doi.org/10.1007/978-3-031-20683-2_5

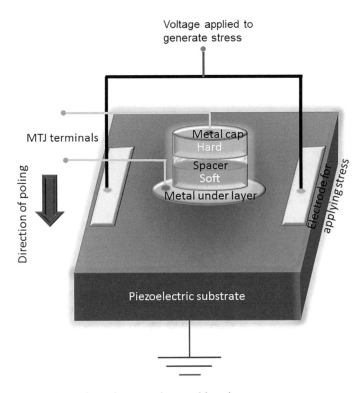

Fig. 5.1 Schematic of a straintronic magnetic tunnel junction

the magnetization of the soft layer will not stay pointing along the hard axis, but relax to the easy axis, thereby erasing the stored bit.

There has been a recent claim of a *non-volatile* straintronic MTJ where dipole interaction between the hard and soft layers supposedly keeps the magnetization of the soft layer pointing along the hard axis even after removal of the strain-generating voltage across the piezoelectric, thereby making the switch appear *non-volatile* [1]. However, this has been questioned and disputed [2]. Dipole interaction normally cannot make this happen and it is likely that what made it happen is that the strain in the piezoelectric did not vanish after removal of the voltage. There was remanent strain (such remanent strain has been frequently reported [3–8]) and that remanent strain (not dipole interaction) kept the magnetization pointing along the hard axis after the voltage was switched off. Unfortunately, that does not make the MTJ reliably non-volatile since the remanent stress can relax spontaneously (e.g., owing to temperature change) thereby erasing the stored bit and making the memory element *volatile* in the end.

One of the earliest demonstrations of a volatile straintronic MTJ was reported in a system shown in Fig. 5.2 where an MTJ with rectangular cross-section was deposited on a (011)-PMN-PT piezoelectric substrate poled in the (011) direction, whose edges had the

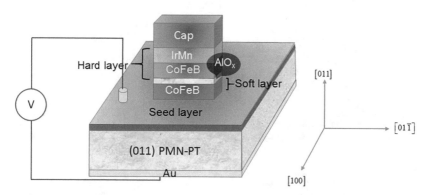

Fig. 5.2 An early straintronic magnetic tunnel junction adapted from *Adv. Mater.*, **26**, 4320 (2014) with permission of Wiley

crystallographic direction depicted in Fig. 5.2 [9]. Application of a positive electric field along the (011) direction generates compressive strain along the [001] direction and tensile strain along the $[01\bar{1}]$ direction and a negative electric field along the same direction will reverse the signs of the strains. From Fig. 5.2, it is clear from the voltage polarity that tensile strain is generated along the short axis of the soft layer and compressive strain along the long axis. Since the soft layer material CoFeB has positive magnetostriction, the $\lambda_s\sigma$ product (with σ being the stress along the long axis) is negative. This will rotate the magnetization of the soft layer from the long axis (easy axis) to the short axis (hard axis), thereby changing the MTJ resistance. Ref. [9] did not directly report the change in the MTJ resistance as a function of the applied electric field generating strain but instead reported the change in the tunneling magnetoresistance (TMR) which is the ratio $(R_{AP} - R_P)/R_P$. The ratio changed by 15% for an electric field of 8 kV/cm.

A later experiment utilized MTJs with much smaller cross-section than the one in ref. [9] and showed remarkable change in the MTJ resistance leading to a TMR almost as large 100% at room temperature [10]. It also demonstrated that strain modulated the coercivity of the soft layer.

The structure of this device is shown in Fig. 5.3, which also shows the strain distribution around the MTJ in the piezoelectric substrate when a voltage is applied across it to generate strain, and the magneto-resistance traces obtained at different applied voltages. Figure 5.4 shows the simulated magnetization distributions within the soft layer at two different gates voltages and also the resistance switching as a function of the applied voltage. The TMR reached ~100% at room temperature, which is an excellent figure. The gate voltage needed to reach this high TMR was 150 V. Since the substrate thickness was 0.5 mm, this translates to an electric field of 3 kV/cm. Therefore, if one can reduce the piezoelectric layer thickness to 1 micron, the gate voltage will drop to 300 mV.

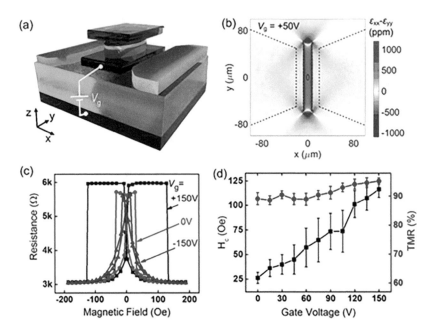

Fig. 5.3 **a** Schematic of a straintronic MTJ with large TMR. A voltage V_g is applied across the piezoelectric layer to generate strain in the soft layer. **b** The in-plane anisotropic strain $\varepsilon_{xx} - \varepsilon_{yy}$ profile generated in the piezoelectric layer by a voltage $V_g = +50$ V. The solid line ellipse at the center denotes the MTJ pillar, and the dashed lines denote the positions of electrodes and side gates shown in (**a**). This result was generated with COMSOL Multiphysics software. **c** Magnetoresistance traces measured under different gate voltages V_g. **d** Variation of the switching (magnetic) field [squares] and tunneling magnetoresistance ratio (TMR) [circles] of the MTJ as a function of V_g. Reproduced from *Appl. Phys. Lett.*, **109** 092, 403 (2016) with the permission of AIP Publishing

5.2   Non-volatile Memory with Mixed Mode (Straintronic + STT) MTJ

MTJs switched with spin transfer torque (STT) are different from straintronic MTJs in that they are usually *non-volatile* since STT can rotate magnetization easily by 180°, unlike strain. Therefore, they form the bedrock of a mainstream magnetic memory technology called "spin-transfer-torque-random-access-memory" (STT-RAM). The MTJ used in STT-RAM is a two-terminal device and the same two terminals can be used to both read and write bits into it. That has an advantage in that it makes the footprint small (which is very important for memory), but it also has some disadvantages like "read disturb", which means that the current passed through the device to measure the resistance (and hence read the bit) can also unintentionally erase or change the bit. Furthermore, using the same current paths to read and write can also reduce the memory's endurance, although that is not a major concern these days. The primary drawback, however, is that

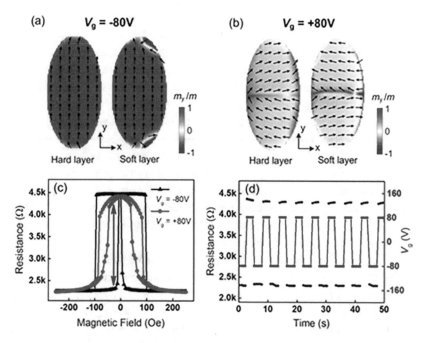

Fig. 5.4 **a**, **b** Micromagnetic simulation results showing the magnetization configurations of the hard and soft layers of the MTJ in Fig. 5.3, after application of gate voltage **a** $V_g = -80$ V and **b** $V_g = +80$ V. A small bias field of 30 Oe is applied along the major axis to overcome any effect of dipole interaction. The dimension of the MTJ is: minor axis = 3 μm and major axis = 6 μm. Black arrows indicate the direction of magnetic moments. **c** Measured magnetoresistance loops for $V_g = -80$ V and $V_g = +80$ V. The blue arrow indicates the switchable high- and low-resistance states. **d** Toggling of the MTJ between high- and low-resistance states with application of ±80 V gate voltage pulsing. A small bias magnetic field of 30 Oe is applied along the +y-axis (refer to Fig. 11a) to overcome the dipole interaction between the two magnetic layers. Reproduced from *Appl. Phys. Lett.*, **109** 092,403 (2016) with the permission of AIP Publishing

the STT-RAM element is extremely energy hungry and the energy dissipated in writing a bit is orders of magnitude larger than that dissipated in switching a field-effect transistor, let alone a straintronic MTJ. This is especially true when the soft layer has to have a reasonably large internal (shape anisotropy) energy barrier for good thermal stability. The current density required to switch is sometimes as large as 10^{10} A/cm^2. As a result, the amount of energy that is dissipated to switch the STT-RAM in ~1 ns could be about 10^7 kT of energy at room temperature (~1.6 pJ), even when the energy barrier within the nanomagnet is only few tens of kT [11]. Further advances have brought this number down to ~100 fJ [12], but that is still excessive since it is three orders of magnitude larger than the energy dissipated in switching a modern day transistor. Needless to say, this would be an enormous price to pay for non-volatility.

Considerable amount of research has now resulted in current densities as low as 2.1 MA cm^{-2} in an MTJ with a resistance-area product of 16 Ω μm^2 [13]. Even with this advance, in an MTJ whose cross-sectional area is 1 μm^2, the power dissipated (I^2R) during the switching action will be ~34 mW, which is extremely high. This is why small cross-sectional areas, much smaller than 1 μm^2, are needed for STT switching. Attempting to reduce the switching current further by thinning the magnetic layers or the spacer layer results in dramatic reduction of the high- to low resistance ratio, or the tunneling magnetoresistance ratio (TMR), since it is governed by spin-dependent tunneling between the magnetic layers. There are well known proposals by researchers to reduce the energy dissipation by using spacer layers that have smaller bandgap, such as ScN, which would offer a lower tunneling resistance and hence a lower resistance-area product, but the lower barrier to tunneling may also increase the thermionic emission over the barrier. Since thermionic emission is not spin-dependent unlike tunneling, increasing it will reduce the TMR and hence the resistance on/off ratio of the MTJ.

In an effort to reduce the energy dissipation associated with the writing of bits in conventional STT-RAM, a *hybrid* writing mechanism that combines STT and straintronics was proposed [14]. It leverages a time-varying strain generated by a surface acoustic wave to aid the switching and this reduces the STT current required to switch. The soft layer of the MTJ is a two-phase multiferroic, i.e. a magnetostrictive layer elastically coupled with an underlying piezoelectric layer. A surface acoustic wave (SAW) is launched in the piezoelectric layer, which strains the soft layer periodically in time. An STT current pulse is passed *synchronously* through the MTJ during the appropriate cycle of the SAW to generate a spin transfer torque that will work in concert with the SAW to orient the magnetization of the soft layer and write the desired bit. The advantage of the hybrid scheme is that a much lower STT current density is needed to write the bit compared to conventional STT-RAM.

The reason why the high current requirement is mitigated is that the SAW rotates the magnetization by 90° during the cycle when the product of the magnetostriction and strain is negative. If *during that cycle*, a current is injected into the MTJ to produce a spin transfer torque (STT), then a complete 180° rotation can be achieved with *reduced current density* since the SAW has already done the heavy lifting and rotated the magnetization by 90°. All that the STT has to do now is to rotate by an additional 90°. Since SAW or straintronics is considerably more energy-efficient than STT, the overall energy dissipation can be reduced by perhaps an order of magnitude [14].

This methodology has its own stringent requirements. One must make sure that the probability of switching is nearly 100% after the STT current is injected and ~0% when no STT current is injected. This is because the SAW flows continuously and if it can switch the magnetization by itself with no help from the STT, then one will get periodic unintentional switching. It was shown in ref. [14] that this stringent requirement can be met with proper design. The structure for this memory cell is shown in Fig. 5.5. A very small amount of energy is required to generate the global SAW that acts on all MTJs on the wafer, and when that energy is amortized over all the MTJs, the energy cost per MTJ is miniscule. It was found that this approach can reduce the write energy dissipation in a memory cell by approximately an order of magnitude. Further reduction may be possible with design optimization.

This idea motivated a recent work which showed that SAW alone (with no STT current present) can modulate the resistance of MTJs whose soft layers are magnetostrictive [15].

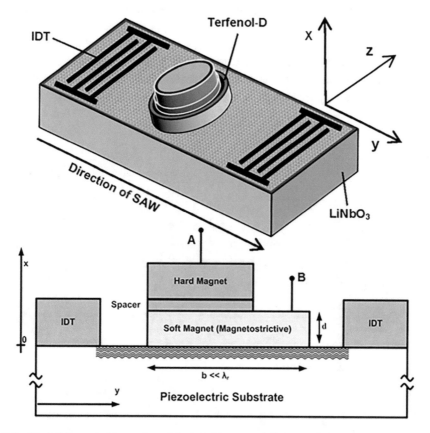

Fig. 5.5 (Top) Schematic illustration of the hybrid system with interdigitated transducers (IDTs) to launch the SAW and an MTJ, serving as a bit storage unit, placed between IDTs on a piezoelectric substrate. The soft layer of the MTJ is in contact with the substrate and is periodically strained by the SAW. The resistance between the terminals A and B is used to read the bit stored (it is assumed that both magnets are metallic). For writing, a small spin polarized current is passed between the same two terminals during the appropriate cycle of the SAW, when the magnetization rotates out of the easy axis. In this configuration, the reading and writing currents do not pass through the highly resistive piezoelectric, so the dissipation during the read/write operation is kept small. Bits are addressed for read/write using the traditional crossbar architecture. Reproduced from *Appl. Phys. Lett.,* **103**, 232,401 (2013) with the permission of AIP Publishing

Figure 5.6a shows the structure of an MTJ fabricated on a LiNbO$_3$ substrate and excited by a SAW launched in the substrate from inter-digitated transducers (IDT). The hard and soft layers are made of CoFeB, which have small positive magnetostriction, and the spacer layer is made of MgO. This structure was chosen because it is known to produce good TMR.

Because of the dipole interaction between the hard and soft layers, the MTJ will prefer to be in the antiparallel state in the absence of any SAW. Figure 5.6b and c show, respectively, the antiparallel resistance of the MTJ as a function of SAW frequency for a fixed SAW amplitude, and as a function of SAW amplitude for a fixed SAW frequency. The dip in Fig. 5.6b occurs very close to the resonant frequency of the IDTs when the launched SAW power is maximum, showing that the SAW (time varying periodic strain) can indeed rotate the magnetization of the soft layer when it is of sufficient power, and thus reduce the antiparallel resistance. Figure 5.6c shows that the higher is the SAW

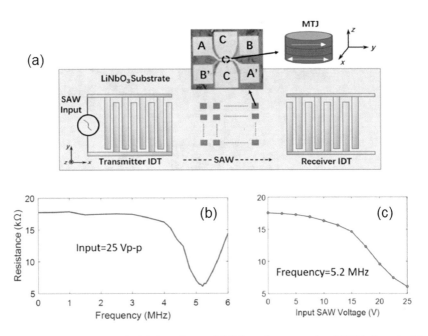

Fig. 5.6 **a** An array of MTJs is fabricated on a LiNbO$_3$ substrate and a SAW is launched in the substrate with interdigitated transducers (IDTs). Each MTJ has six electrodes as shown in the inset; either A-B or A'-B' electrodes were used to measure the MTJ resistance while the electrode C was not used. **b** Measured antiparallel resistance as a function of SAW frequency for a fixed peak-to-peak SAW voltage amplitude. The dip occurs close to the resonant frequency of the IDTs where the launched SAW power is maximum for that voltage amplitude. **c** Measured antiparallel resistance as a function of SAW voltage at the frequency where the dip occurs in Fig. 5.6b. Reproduced from *J. Appl. Phys.*, 130, 033,901 (2021) with the permission of AIP Publishing

power, the more is the rotation and more is the decrease in the antiparallel resistance. These studies confirm that a SAW can affect the resistance state of an MTJ appreciably.

5.3   Straintronic Ternary Content Addressable Memory

Straintronics can also be exploited to implement unconventional memory such as *ternary content addressable memory* (TCAM) with vastly improved properties. These implementations are not just very energy-efficient, but they also have much smaller footprints than conventional implementations with CMOS transistors.

In ternary content-addressable memory (TCAM), a memory cell is searched by content rather than row and column address. A TCAM compares input search data against a table of stored data and then returns the memory address of entirely or partially matching data. In each TCAM cell, the search and storage bits have three states: "0," "1," and "X" (don't care). The "don't care" state allows masking, i.e., a match is indicated no matter what the storage and search bits are.

A TCAM cell can be implemented very effectively with a *skewed* straintronic MTJ shown in Fig. 5.7a. This is an MTJ in which the easy axes of the hard and soft layers are *non-collinear*. In such an MTJ, the magnetizations of the hard and soft layer will be at an *obtuse angle* because dipole coupling between the two layers will always tend to maximize the angle between them. Figure 5.7b shows the top view of a skewed MTJ where the misalignment between the easy axes of the two ferromagnetic layers is made 45° and hence the angle between the magnetizations of the two layers in the absence of strain is 135°.

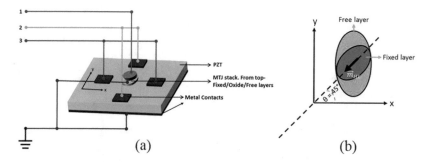

Fig. 5.7 **a** Schematic of a skewed straintronic magneto-tunneling junction with four electrodes. The antipodal pairs are electrically shorted and designated as terminals 2 and 3. Voltages are applied between these terminals and ground to strain the MTJ. The resistance of the MTJ is measured between terminal 1 and ground; **b** top view of the hard (fixed) and soft (free) layers of the skewed MTJ showing that the angle between the magnetizations of the two layers is 135° in the absence of any strain. Reproduced from *IEEE Trans. Elec. Dev.*, **64,** 2835 (2017) with permission of the Institute of Electrical and Electronics Engineers

When strain is applied to the skewed MTJ by applying a voltage of appropriate polarity to terminal 2, the magnetization of the soft layer will rotate *clockwise* (not anti-clockwise) [in Fig. 5.7b] because dipole coupling always tends to maximize the angle between the magnetizations of the two layers. Therefore, the angular separation between the magnetizations of the two layers will start out at 135°, go through 180° and stop at 225° (equivalent to 135°) after completing the maximum allowed 90° rotation. Since the MTJ resistance is $R_{MTJ} = R_P + \frac{R_{AP}-R_P}{2}[1 - \cos\theta]$, with θ being the angle between the magnetizations of the two layers, the MTJ resistance will start out as $(R_{AP} + R_P)/2 + (R_{AP} - R_P)/(2\sqrt{2})$, peak at R_{AP}, and then decrease back to $(R_{AP} + R_P)/2 + (R_{AP} - R_P)/(2\sqrt{2})$. If one applies a constant voltage at terminal 1, then the current I_{1G} at terminal 1 (which is the current through the MTJ) as a function of the voltage V_{2G} applied to terminal 2, will exhibit a "notch" which will occur when the MTJ resistance reaches its peak (becomes R_{AP}). Figure 5.8 shows the angular separation between the magnetizations of the hard and soft layers, and the MTJ current I_{1G} as functions of the voltage V_{2G} simulated by solving the stochastic Landau-Lifshitz-Gilbert equation. The results are shown for two different temperatures: 0 K (no thermal noise present) and 300 K (thermal noise is present). This notch characteristic is extremely unusual and will be very difficult to implement with field-effect transistors (impossible to implement with a single transistor). This highlights some of the extra-ordinary attributes of straintronics.

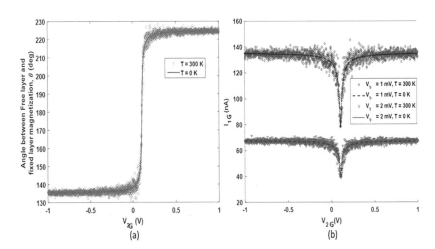

Fig. 5.8 **a** The angle between the magnetizations of the hard and soft layers as a function of the voltage V_{2G} applied between terminal 2 and ground; **b** the current I_{1G} flowing through the MTJ (for a constant voltage applied at terminal 1) as a function of V_{2G}. The results are shown for 0 K and 300 K temperatures. Reproduced from *IEEE Trans. Elec. Dev.*, **64**, 2835 (2017) with permission of the Institute of Electrical and Electronics Engineers

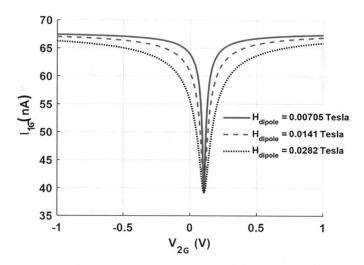

Fig. 5.9 Transfer characteristic of a skewed straintronic MTJ as a function of the dipole coupling strength between the hard and soft layers at 0 K temperature. Reproduced from *IEEE Trans. Elec. Dev.*, **64**, 2835 (2017) with permission of the Institute of Electrical and Electronics Engineers

The *sharpness* of the notch in the transfer characteristic shown in Fig. 5.8b depends on the strength of dipole coupling between the hard and soft layers. The effect of dipole coupling on the soft layer can be approximated by an effective magnetic field along the easy axis of the hard layer that is directed opposite to the magnetization of the hard layer. Figure 5.9 shows the transfer characteristic as a function of dipole coupling strength. Weaker dipole coupling makes the notch sharper.

The position of the notch can be shifted along the V_{2G} axis by applying a voltage V_{3G} between terminal 3 and ground, which changes the value of the strain generated at any given voltage V_{2G}. The voltage V_{2G} where the notch occurs is related to V_{3G} approximately as $V_{2G}^{\text{notch}} = V_{3G} - V_F$, where V_F is a fixed voltage. Figure 5.10 shows the transfer characteristics (I_{1G} versus V_{2G}) for three different values of V_{3G}.

In the TCAM operation, the search bits are encoded in the potential V_{2G} and the stored bits in the potential V_{3G}. Let us say that the search bits X, 0 and 1 are encoded in voltages -0.5, -0.05 and $+0.1$ V, respectively. The store bits 1, 0 and X are encoded in those values of V_{3G} that will place the centers of the notches in the transfer characteristics at -0.05 V, $+0.1$ V and $+0.25$ V, respectively as shown in Fig. 5.10. That means these values of V_{3G} are approximately $-0.05 + V_F$, $0.1 + V_F$ and $0.25 + V_F$ volts. In the simulated system, $V_F = -0.1$ V and hence the three values of V_{3G} are -0.15, 0 and 0.15 V.

In the encoding scheme, a high current I_{1G} denotes a match between the stored and search bits.

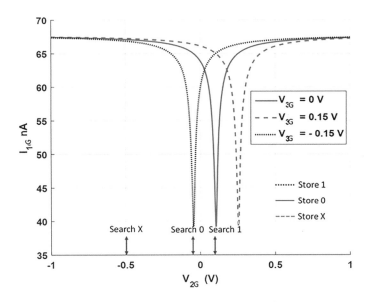

Fig. 5.10 The current I_{1G} as a function of the voltage V_{2G} for three different values of V_{3G}. Reproduced from *IEEE Trans. Elec. Dev.*, **64,** 2835 (2017) with permission of the Institute of Electrical and Electronics Engineers

When the stored bit is 1 and the search bit is 0, $V_{3G} = -0.15$ V and V_{2G} is -0.05 V. Looking at Fig. 5.10, we find that we are located in a notch (the far left notch in Fig. 5.10) so that the current through the MTJ is *low* and we have the correct "no-match" result. Similarly, when the search bit is 1 and the stored bit is 0, $V_{3G} = 0$ V and $V_{2G} = 0.1$ V, so that we are located in the center notch and the current is again low indicating no match. When the search and stored bits are the same, we are clearly not in a notch (see Fig. 5.10), so the current through the MTJ is high, and the match is correctly indicated.

Let us now examine what happens with the "don't care" bit. Since the notch for the stored bit X (corresponding to $V_{3G} = 0.15$ V) is located farthest to the right and exceeds all three voltages V_{2G} encoding the three search bits 0, 1 and X, the current I_{1G} remains high for all search bits, indicating a match no matter what the search bit is, as long as the stored bit is X. Finally, the search bit X is encoded in a voltage of $- 0.5$ V which is to the left of all notches. Hence, when the search bit is X, no matter what the stored bit is, we are never in a notch and the current is always high, indicating a match. Thus, the correct TCAM operation is realized.

If one had tried to implement a static TCAM cell with CMOS transistors, one would have needed 16 transistors [17]. Here, one can get by with just a single MTJ, which reduces the circuit footprint dramatically. Reference [16] examined a large number of TCAM based circuitry realized with skewed straintronic MTJs and found significant reduction in energy and increase in speed as well.

5.4   Memory Scaling Issues in Straintronic Memory

This chapter is concluded with the discussion of a very import scaling issue that afflicts straintronic memory of all kinds. Memory requires high density (in fact it is the most important commercial consideration) which would demand that straintronic memory be implemented with *perpendicular* magnetic tunnel junctions (p-MTJs) that can have lateral dimensions less than 20 nm. Larger lateral dimensions may not be competitive with experimentally demonstrated STT-RAM of 11 nm diameter [18].

The reason why nanomagnets with perpendicular magnetic anisotropy can have much smaller volumes and hence much smaller lateral dimensions than those with in-plane magnetic anisotropy is well-known. One must keep the energy barrier within the nanomagnet large enough for good thermal stability and long memory retention times. The energy barrier is $E_b = K_u \Omega$, where K_u is the uniaxial anisotropy energy density and Ω is the nanomagnet's volume. The value of K_u is much larger for perpendicular magnetic anisotropy (PMA) than for in-plane anisotropy (IMA) and hence one can obtain the same energy barrier in a smaller volume with perpendicular anisotropy compared to in-plane anisotropy. This is why p-MTJs can have smaller footprints than MTJs possessing in-plane anisotropy, for the same energy barrier and thermal stability.

For a soft layer of diameter ~20 nm and thickness ~1 nm, which is typical for p-MTJs, the volume is ~314.2 nm^3. Assume that we need E_b~1 eV = 40 kT which will ensure thermal stability and long retention times. Thus, we will need $K_u = 5.1 \times 10^5$ J/m^3.

To switch the magnetization of the soft layer with strain alone, one will need to overcome the energy barrier E_b with the strain anisotropy energy and hence need that $K_u < (3/2)|\lambda_s \sigma|$. Even if we assume an optimistic $|(3/2)\lambda_s| \sim 500$ ppm, the stress $|\sigma|$ required will be ~1000 MPa, which is very difficult to apply either via direct strain transferred from an underlying piezoelectric layer or by the use of surface acoustic waves (SAW). Highly magnetostrictive materials with large λ_s, for example Terfenol-D [19], will not help because of the bidirectional coupling between the magnetization and strain, which reduces the strain anisotropy energy [18]. Hence, straintronic p-MTJs are difficult to implement and therefore most demonstrated straintronic switching has employed nanomagnets with in-plane anisotropy. An additional problem with straintronic p-MTJs is that their switching times will be much longer than those with in-plane anisotropy. In nanomagnets with in-plane magnetic anisotropy, the out of plane excursion of the magnetization vector (while switching under strain) speeds up the switching by generating an additional torque on the magnetization. This effect is absent in p-MTJs built with nanomagnets possessing perpendicular magnetic anisotropy, which is why straintronic p-MTJs will be much more sluggish in switching. That is why almost all straintronic MTJs have in-plane magnetic anisotropy and the ones we have considered in this monograph also have in-plane anisotropy.

This brings us to an impasse. The overriding consideration in memory is neither the energy dissipation nor the speed, but the density, from a commercial perspective. In fact, this is the reason why SOT memory has not displaced STT memory commercially (despite the former's lower energy dissipation) since SOT switches typically need three terminals while STT memory needs only two. That makes the density of SOT memory considerably lower than that of STT memory. *Similar considerations would compel us to ultimately explore straintronic p-MTJs in spite of all the other disadvantages.*

To build straintronic p-MTJs, one has to first contend with the K_u problem that was discussed earlier. It might appear that one could get away with a lower K_u for a given energy barrier E_b (and hence lower stress needed to switch) if one were to increase the nanomagnet volume, since $K_u = E_b / \Omega$. . One might be tempted to do this by increasing the thickness of the nanomagnet while keeping the lateral footprint the same (so as to not reduce the areal density), but this does not work since increasing the thickness results in loss of perpendicular magnetic anisotropy (PMA), which is typically quenched when the nanomagnet thickness exceeds ~1 nm. Thus, straintronic p-MTJs remain elusive. Today, this is the most serious challenge to straintronic memory.

References

1. A. Chen, H.-G. Piao, M. Ji, B. Fang, Y. Wen, Y. Ma, P. Li, X.-X. Zhang, Adv. Mater. 2105902 (2021)
2. S. Bandyopadhyay, Comment on 'Using dipole interaction to achieve nonvolatile voltage control of magnetism in multiferroic heterostructures. Adv. Mater. 2105902 (2021), arXiv:2203.08236.
3. L. Yang, Y. Zhao, S. Zhang, P. Li, Y. Gao, Y. Yang, H. Hu, P. Miao, Y. Liu, A. Chen, C.W. Nan, C. Gao, Bipolar loop-like non-volatile strain in the (001)-oriented $Pb(Mg_{1/3}Nb_{2/3})$-$PbTiO_3$ single crystals. Sci. Rep. **4**, 4591 (2014)
4. L. Yu, S.W. Yu, X.Q. Feng, Effects of electric fatigue on the butterfly curves of ferroelectric ceramics. Mater. Sci. Eng. A **459**, 273 (2007)
5. D.C. Lupascu, C. Verdier, Fatigue anisotropy in lead-zirconate-titanate. J. Eur. Ceram. Soc. **24**, 1663 (2004)
6. T. Wu et al., Domain engineered switchable strain states in ferroelectric (011) $[Pb(Mg_{1/3}Nb_{2/3})O_3]_{(1-x)}$-$[PbTiO_3]_x$ (PMN-PT, x≈0.32) single crystals. J. Appl Phys. **109**, 124101 (2011)
7. T. Wu et al., Electrical control of reversible and permanent magnetization reorientation for magnetoelectric memory devices. Appl. Phys. Lett. **98**, 262504 (2011)
8. M. Liu, B.M. Howe, L. Grazulis, K. Mahalingam, T. Nan, N.X. Sun, G.J. Brown, Voltage-impulse-induced non-volatile ferroelastic switching of ferromagnetic resonance for reconfigurable magnetoelectric microwave devices. Adv. Mater. **25**, 4886 (2013)
9. P. Li, A. Chen, Y. Zhao, S. Zhang, L. Yang, Y. Liu, M. Zhu, H. Zhang, X. Han, Electric field manipulation of magnetization rotation and tunneling magnetoresistance of magnetic tunnel junctions at room temperature. Adv. Mater. **26**, 4320 (2014)
10. Z.Y. Zhao, M. Jamali, N. D'Souza, D. Zhang, S. Bandyopadhyay, J. Atulasimha, J.-P. Wang, Giant voltage manipulation of MgO-based magnetic tunnel junctions via localized anisotropic strain: a potential pathway to ultra-energy-efficient memory technology. Appl. Phys. Lett. **109**, 092403 (2016)

11. K.L. Wang, P. Khalili-Amiri, Non-volatile spintronics: Perspectives on instant on non-volatile nanoelectronic systems. SPIN **2**, 1250009 (2012)

12. K.L. Wang, J.G. Alzate, P. Khalili-Amiri, Low-power non-volatile spintronic memory: STT-RAM and beyond. J. Phys. D: Appl. Phys. **46**, 074003 (2013)

13. H. Meng, R. Sbiaa, S.Y.H. Lua, C.C. Wang, M.A.K. Akhtar, S.K. Wong, P. Luo, C.J.P. Carlberg, K.S.A. Ang, Low current density induced spin-transfer torque switching in CoFeB–MgO magnetic tunnel junctions with perpendicular anisotropy. J. Phys. D: Appl. Phys. **44**, 405001 (2011)

14. A.K. Biswas, S. Bandyopadhyay, J. Atulasimha, Acoustically assisted spin-transfer-torque switching of nanomagnets: an energy-efficient hybrid writing scheme for non-volatile memory. Appl. Phys. Lett. **103**, 232401 (2013)

15. D. Bhattacharya, P. Sheng, M.A. Abeed, Z. Zhao, H. Li, J.-P. Wang, S. Bandyopadhyay, B. Ma, J. Atulasimha, Surface acoustic wave induced modulation of tunneling magnetoresistance in magnetic tunnel junctions. J. Appl. Phys. **130**, 033901 (2021)

16. S. Dey Manasi, M.M. Al-Rashid, J. Atulasimha, S. Bandyopadhyay, A.R. Trivedi, Straintronic magneto-tunneling junction based ternary content-addressable memory (Parts I and II). IEEE Trans. Elec. Dev. **64**, 2835 (2017)

17. J.J. Nowak, R.P. Robertazzi, J.Z. Sun, G. Hu, J.H. Park, J. Lee, A.J. Annunziata, G.P. Lauer, R. Kothandaraman, E.J.O. Sullivan, P.L. Trouilloud, Y. Kim, D.C. Worledge, Dependence of voltage and size on write error rates in spin-transfer torque magnetic random-access memory. IEEE Magn. Lett. **7**, 1 (2016)

18. L. Sandlund, M. Fahlander, T. Cedell, A. E. Clark, J. B. Restorff, M. Wun-Fogle, Magnetostriction, elastic moduli, and coupling factors of composite Terfenol-D. J. Appl. Phys. **75**, 5656 (1994)

19. Z. Xiao, R. Lo Conte, C. Chen, C.Y. Liang, A. Sepulveda, J. Bokor, G.P. Carman, R.N. Candler, Bi-directional coupling in strain-mediated multiferroic heterostructures with magnetic domains and domain wall motion. Sci. Rep. 8, 5207 (2018)

Straintronic Boolean Logic: Energy-Efficient but Error-Prone

6

One of the earliest explored application of straintronics was in Boolean logic circuitry since Boolean logic has always been the backbone of mainstream digital computing and signal processing. There are primarily two components that have to be implemented for Boolean logic: a universal Boolean logic gate (NAND or NOR) and a connection scheme that will steer digital bits *undirectionally* from one gate to the next. Unidirectionality is required to ensure that it is the input logic variable which determines the state of the output logic variable and not the other way around (in other words, a master-slave relationship must exist between the input and the output). This feature is sometimes referred to as *isolation between input and output.* Unfortunately, what is sometimes not appreciated is that having these two components are necessary, but not sufficient, conditions for building Boolean logic circuits.

There are actually six requirements for logic sufficiency [1–3]: (1) *Concatenability*: the input and output logic bits must be encoded in the same physical quantity so that the output of one logic gate can be fed *directly* to the input of the next logic gate, without requiring an intervening "transducer" to convert the output of the preceding stage to a quantity that can become the input of the succeeding stage. For example, if the input is encoded in voltage states, then the output must also be encoded in voltage states in order to be fed directly to the input(s) of the next logic gate. If the input(s) and output(s) are different quantities (e.g. the input bit is encoded in voltage and the output bit is encoded in the direction of a magnetic field), then a suitable transducer will be required between two consecutive stages to convert the direction of a magnetic field into a voltage level. That transducer may have a much larger footprint than the gate itself and may also consume orders of magnitude more energy than the gate. It is therefore imperative to not have *any transducer,* which is why gates must be concatenable. (2) *Non-linearity and gain*: these are required for logic level restoration in noisy environments where the distinction between the two bits 0 and 1 can get blurred as a bit propagates through a logic chain.

S. Bandyopadhyay, *Magnetic Straintronics*, Synthesis Lectures on Engineering, Science, and Technology, https://doi.org/10.1007/978-3-031-20683-2_6

To understand this, assume that logic bit 0 is encoded in voltage between 0 and 1 V and logic bit 1 is encoded in voltage between 3 and 4 V. However, because of noise, the bit 0 has become 1.5 V and bit 1 has become 2.5 V, which shrinks the separation between the logic levels to 1 V instead of $3 - 1 = 2$ V. Now, if the logic device has a gain of 2, then the 1.5 V will be amplified to 3 V and the 2.5 V will be amplified to 5 V so that the separation once again becomes $5 - 3 = 2$ V. This is called logic level restoration. (3) *Isolation between input and output*: The input should affect the output and not the other way around. This will ensure than logic signal always flows unidirectionally from the input stage to the output stage. (4) *Universal gate operation:* A universal gate ensures that any Boolean function can be implemented with the gate and any arbitrary combinational or sequential logic circuit can be built. (5) *Reliability:* The error probability must be very low since errors in logic circuits *propagate*. If the output of one gate is corrupted and is fed as input to a succeeding gate, then the output of the second gate is also corrupted, and so on. This is very much unlike memory, where if one stored bit is corrupted, it does not infect any other bit. (6) *Scalability*: There should be no serious and fundamental impediment to downscaling in size.

Unfortunately, many proposals and ideas for magnetic Boolean logic circuits (built with various types of magnetic devices and elements) that are found in the literature do not satisfy one or more of these requirements, which calls into question their viability and validity [4].

6.1   Dipole Coupled Nanomagnetic Logic

One of the earliest ideas for implementing very low power Boolean logic with nano-magnets is a paradigm termed "magnetic quantum cellular automata" [5] where dipole coupling between neighboring nanomagnets was leveraged to steer bits, encoded in the bistable magnetization states of nanomagnets, from one nanomagnet to another.[1] A universal Boolean gate operation was realized by manipulating the magnetic states of the nanomagnets with a specific signaling scheme. Unfortunately, the input and output variables were dissimilar quantities which made the gates non-concatenable and hence inappropriate for Boolean logic. However, the lure of using dipole interaction between nanomagnets to act as "virtual wires" that consume no area and dissipate no energy (because dipole interaction does not involve current flow) was very enticing. Similar ideas were proposed earlier where the quantum mechanical spins of single electrons in quantum dots subjected to a global magnetic field encoded binary bits and exchange interaction between nearest neighbors was manipulated to elicit Boolean logic functionality [6, 7]. The difference between the two schemes is that the magnets act as a giant classical spin which is much more robust against noise and thermal perturbations than a single electron

---

[1] This scheme, despite its name, was all about Boolean logic and had nothing to do with "cellular automata" which is an entirely different paradigm for digital computing.

spin, and dipole interaction decays much more slowly with increasing separation between the interacting elements than exchange interaction.

A concatenable version of magnetic quantum cellular automata was later demonstrated where a majority logic gate was realized, but only one out of four gates worked, showing that the error probability was 75%, making such constructs unreliable [8]. The high error rate probably accrued from misalignment between nanomagnets which can have lethal consequences in such constructs [9]. A later experiment used PMA nanomagnets to build the majority logic gate with the hope that the reliability will improve because the magnetization dynamics of PMA nanomagnets' are relatively immune to shape and size variations and maybe even misalignment [10]. However, even though a low error rate was claimed, it was not mentioned what this error rate was. Another paper reported "error-free" propagation of bits in a chain of nanomagnets implementing a dipole coupled inverter [11], but there was insufficient statistics and details to estimate the error probability. Suffice it to say then that we do not know how reliable dipole coupled logic can be, based on published experimental results. In 1956, John von-Neumann had shown that the *maximum* tolerable error probability in a single majority logic gate working in isolation is about 0.0073 [12] and it will be several orders of magnitude smaller than that when the gate is used in circuits containing tens or hundreds of millions of gates. In fact, there are some indirect claims that the maximum tolerable gate error probability in modern logic circuits may be as low as 10^{-15} [13] which is several orders of magnitude smaller than what has been achieved or even seems possible in dipole coupled nanomagnetic logic.

6.2 Straintronic Dipole Coupled Nanomagnetic Logic

In spite of the impractically high error rates, the allure of having a "wireless" architecture for Boolean logic, which eliminates the dissipation associated with current flow through the wires between logic stages, had prompted research in *straintronic* implementation of the two major components required for rudimentary dipole-coupled nanomagnetic Boolean logic: a universal straintronic logic gate, and a straintronic way for steering bits, encoded in the bistable magnetization states of nanomagnets, unidirectionally from one logic gate to the next. The straintronic dipole coupled magnetic logic gates are concatenable since the inputs and outputs are identical variables, namely magnetization orientation. Logic level restoration is automatic since the magnetization orientations are bistable and can have only two directions (unlike voltage or current encoding binary variables; voltage and current are analog quantities). Unidirectionality is enforced by a scheme called *Bennett clocking* which is discussed later. However, reliability and scalability are poor. The scalability is poor since dipole coupling strength varies as the square of the nanomagnet volume and decreasing the volume will weaken dipole interaction strength. This is a shortcoming of all dipole coupled magnetic logic ideas. However, the worse shortcoming is the very poor reliability, which we discuss at length later. Unless some radical improvements can be made in this regard using completely novel approaches, dipole coupled magnetic logic of any kind may be unrealistic.

6.2.1   Straintronic Nanomagnetic Inverter

Before we discuss any universal straintronic dipole coupled magnetic Boolean gate (e.g. a NAND gate), let us first discuss the simplest Boolean logic gate, which is the inverter or NOT gate. This is a one-input and one-output gate, which is not a universal gate, but we discuss it to highlight some basic principles involved in implementing Boolean logic gates with straintronics.

The straintronic implementation of the inverter is shown in Fig. 6.1a. There are two elliptical nanomagnetics, one with more eccentricity than the other, placed side by side in close proximity to allow for sufficiently strong dipole coupling between them. The input bit (0 or 1) is encoded in the bistable magnetization orientation of the more eccentric nanomagnet and the output bit is encoded in the bistable magnetization of the other nanomagnet. The line joining the centers of the two nanomagnets is collinear with their minor axis (hard axis). In this case, dipole coupling will ensure that in the ground state of this pair, the magnetizations in the two nanomagnets are mutually *antiparallel*. This immediately realizes the NOT gate since the output bit will be the logic complement of the input bit, as long as it can be ensured that the system is in the *ground state*.

The system may not always automatically relax to the ground state after the arrival of a new input that changes the magnetization state of the input nanomagnet. It may get stuck in a metastable state and thereafter not be able to reach the ground state. To illustrate this case, consider that we flipped the input bit, thereby placing the system temporarily in the state shown in Fig. 6.1b. One would expect that the output bit will automatically flip in response to the input to attain the antiferromagnetic order (i.e., reach the ground state), but this may not happen because of the shape anisotropy energy barrier within the output nanomagnet which can prevent its magnetization from flipping in response to the changed input magnetization. Dipole coupling energy has to overcome this shape anisotropy energy barrier to make the output magnet's magnetization flip in response to the input magnet's state, but if the dipole coupling is not strong enough, then this will not happen. To understand this, consider the potential energy profile in the plane of the output nanomagnet shown in Fig. 6.1c as a function of the magnetization orientation of the output nanomagnet, (θ is the angle subtended by the magnetization $\vec{M}$ of the output nanomagnet with the easy axis). The potential profile is asymmetric in this case (it is not a double well potential) because of the dipole coupling. The ground state is at $\theta = 180°$, where the magnetization of the output nanomagnet will be antiparallel to that of the input nanomagnet, but the output nanomagnet's magnetization will remain stuck at its original location $\theta = 0°$, which is a local energy minimum (metastable state) and it cannot reach the global minimum (ground state) because of the intervening energy barrier, as shown in

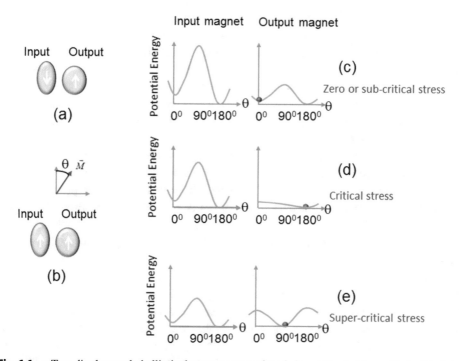

Fig. 6.1 **a** Two dipole coupled elliptical nanomagnets placed close to each other with the line joining their centers lying along the minor axis. The magnetization of the more eccentric nanomagnet
encodes the input bit and that of the less eccentric one the output bit. **b** When the input bit is changed,
the system goes into a metastable state since the output bit may not flip in response to the changed
input bit. This happens because dipole coupling, which would enforce the antiferromagnetic ordering, cannot overcome the shape anisotropy energy barrier within the output nanomagnet to make
its magnetization flip. **c** Potential energy profiles within the two nanomagnets in the absence of
any stress or with sub-critical stress when the magnetizations are in the metastable state shown in
Fig. 6.1(b). The profile is asymmetric because of dipole interaction which makes one magnetization orientation preferred over the other and hence one minimum slightly lower than the other. **d**
Potential energy profiles in the presence of critical stress when the energy barrier within the output
nanomagnet is just eroded, but not inverted. In this case, the output magnetization will flip with high
likelihood to ensure successful NOT operation. **e** Potential energy profiles in the presence of super-
critical stress when the potential barrier within the output nanomagnet is inverted, placing the output
magnetization pointing along the hard axis of the output nanomagnet. Subsequent removal of stress
will place the magnetization temporarily in the maximally unstable state, from which it will decay
to either the metastable state or the ground state with slightly higher preference for the latter owing
to the dipole coupling with the input nanomagnet

Fig. 6.1c. Something will be needed to depress the energy barrier and allow the system
to escape from the metastable state and reach the ground state and implement the NOT
gate operation.

Global stress will depress the energy barrier in both nanomagnets. One will apply
enough stress to erode or invert the potential barrier in the output nanomagnet, but this

stress will not be enough to erode the potential barrier in the input nanomagnet since it has a much higher shape anisotropy energy barrier owing to its much larger eccentricity (this is the reason why it is made more eccentric). "Critical stress" is defined as the stress that will just completely erode, but not invert, the energy barrier in the output nanomagnet, as shown in Fig. 6.1d. The value of this stress $\sigma_{critical}$ is roughly obtained by equating the stress anisotropy energy to the shape anisotropy energy barrier E_b in the output nanomagnet: $(3/2)\lambda_s\sigma_{critical}\Omega = E_b$ where Ω is the nanomagnet volume and λ_s is the saturation magnetostriction.

If one applies critical stress, the state of the system (denoted by the yellow ball in Fig. 6.1d) will easily roll down the potential slope from the initial metastable state at $\theta = 0°$ to the ground state at $\theta = 180°$ with very high probability. The probability will approach 100% at 0 K temperature, if the stress is kept on for a duration longer than it takes to roll down. This results in large likelihood of successful NOT gate operation. Even at room temperature, the probability of successful NOT operation will be much larger than 50%.

If instead one applies super-critical stress, as shown in Fig. 6.1e, then that will *invert* the barrier, placing the output nanomagnet's magnetization along its hard axis temporarily (as long as the stress is on). Subsequently, when one withdraws the stress, the potential profile will be restored to that in Fig. 6.1c with the system state (yellow ball) temporarily perched on top of the energy barrier (the maximally unstable position). From there, the yellow ball will roll down to either of the two minima at $\theta = 0°$ and $\theta = 180°$, but with slightly higher preference for the latter since this minimum is slightly lower than the other due to dipole coupling with the input nanomagnet. In this case, the probability of successful NOT operation will be slightly larger than 50% (nowhere near as large as in the case of critical stress).

What the previous discussion makes clear is that the NOT operation will not succeed without the stress cycle, not to mention the fact that the success probability is very sensitive to how much stress is applied. Hence, stress acts as a "clock" to trigger the gate operation. In dipole coupled straintronic Boolean gates, the role of the stress (or the strain) is usually to act as the *clock*.

6.2.2 Experimental Demonstration of Straintronic Nanomagnetic Inverter

To demonstrate the NOT gate operation, two nanomagnets of Co were fabricated on a piezoelectric substrate [14]. The scanning electron micrographs of the two are shown in Fig. 6.2a. Their magnetizations were aligned parallel to each other with an external magnetic field to mimic the configuration in Fig. 6.1b. Super-critical stress was applied to temporarily orient the output magnetization vector along the minor axis of the output nanomagnet and then stress was relaxed to complete the NOT operation. The sequence is shown in Fig. 6.2b.

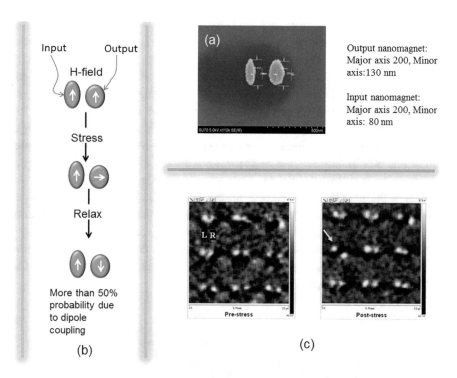

Fig. 6.2 **a** Scanning electron micrographs of two dipole coupled Co nanomagnets fabricated on a PMN-PT piezoelectric substrate. The more eccentric one will be the host for the input bit and the other the host for the output bit. **b** Sequence of operations to demonstrate inverter functionality. **c** Magnetic force micrographs of 9 dipole coupled pairs after their magnetizations have all been aligned in the same direction with a magnetic field. The left panel shows the pre-stress magnetization states (all aligned in the same direction) and the right panel shows the post stress condition. Stress triggered the NOT operation in only one out of the nine pairs shown. Reproduced from *Nano Letters*, **16**, 1069 (2016) with permission of the American Chemical Society

Figure 6.2c shows the magnetic force micrographs depicting the magnetic states of nine dipole coupled pairs before stress application when all of their magnetizations have been aligned in the same direction with a magnetic field (the magnetic field is removed after the alignment is complete). The magnetic states after stress application and subsequent withdrawal are also shown. One would expect at least four of these pairs to exhibit the NOT operation since the probability of success is theoretically greater than 50%, but only one pair (denoted by the yellow arrow) flips indicating that the probability is much less than 50%, contrary to our theoretical expectation. There was not enough stochastic sampling to determine what the exact probability was, but it was surely much less than 50%. This is a consequence of the fact that straintronic logic is error-prone by nature. We will look at the sources of switching errors in the next chapter, but one major contributor happens to be the edge roughness which introduces pinning sites and other imperfections which hinder magnetization reversal. It is obvious from Fig. 6.2a that there is considerable edge roughness in the nanomagnets and this turns out to be major contributor to switching failure.

The initial guess for the source of large error rate in ref. [14] was the weak saturation magnetostriction of Co. It was thought that low magnetostriction (~60 ppm) could not generate enough stress anisotropy energy to overcome the shape anisotropy energy barrier in the output nanomagnets. Accordingly, the experiment was repeated with FeGa nanomagnets which have 5 times larger magnetostriction than Co [15]. The results are shown in Fig. 6.3. In this case, the magnetizations did not even necessarily line up along the major axis (easy axis) since FeGa is an alloy unlike Co, and there are pinning sites that can trap the magnetization away from the major axis. In any case, when two sets of four pairs were examined, one pair seemed to exhibit the NOT behavior in each set. Thus, the error probability did not improve significantly, despite the five times higher magnetostriction. There may be other factors at play here because FeGa is an alloy and has additional pinning sites as a result of material imperfections, which may decrease the probability of successful operation.

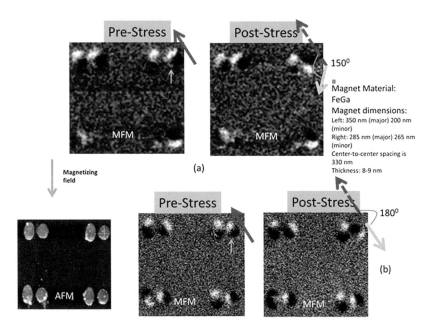

Fig. 6.3 Pre-stress and post-stress magnetic force micrographs of four dipole coupled pairs of FeGa nanomagnets acting as inverters. Not all nanomagnets have good magnetic contrast. **a** In one set, the magnetization of the output nanomagnet rotated by 150° in one out of four pairs and **b** in another set, the magnetization of the output nanomagnet rotated by 180°in one out of four pairs. There is not enough statistics to determine the actual error probability, but it is certainly much higher than 50%. Reproduced from Nanotechnology, 26, 401,001 (2015) with permission of the Institute of Physics

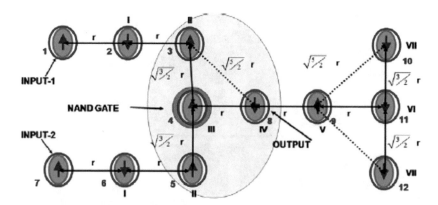

Fig. 6.4 A straintronic NAND gate with fan out where the gate operation is triggered by a 4-phase clock generating strain. The relative center-to-center separations between nanomagnets are shown. The magnetization orientations shown correspond to the correct initial ground states when INPUT-1 = 1 and INPUT-2 = 1. Reproduced from *Nanotechnology*, **23**, 105,201 (2012) with permission of the Institute of Physics

6.2.3   Straintronic NAND Gate with Fan-Out

The NAND gate is a two-input and one-output logic gate that is universal. Reference [16] presented the design of such a gate implemented with dipole coupled elliptical magnetostrictive nanomagnets, where the gate operation was triggered with a four-phase clock generating strain in the nanomagnets. The gate also had fan-out whereby the output could be fed to the inputs of multiple succeeding stages. The schematic of the gate is shown in Fig. 6.4.

Reference [14] showed that a pipelined bit throughput rate of 0.5 GHz is possible with sinusoidal 4-phase clocking. The gate operation can be completed in 2 ns with a latency of 4 ns. The energy dissipation in the gate itself was computed as ~500 kT and that in the peripheral elements as 1250 kT at room temperature.

6.2.4   Steering Logic Bits Unidirectionally from One Logic Stage to the Next: Bennett Clocking with Straintronics

In dipole-coupled logic, a binary bit encoded in the (bistable) magnetization orientation of a nanomagnet, can be propagated *unidirectionally* along a chain of nanomagnets using "Bennett clocking" [17]. The clocking can be done with voltage generated strain [18] and the idea is illustrated in Fig. 6.5.

Consider a row of elliptical magnetostrictive nanomagnets with nearest neighbor dipole coupling. The nanomagnets are placed such that their minor axes (hard axes) lie along the

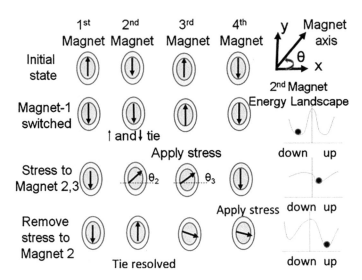

Fig. 6.5 Bennett clocking in a row of nanomagnets to propagate a bit encoded in the magnetization orientation unidirectionally from left to right along the chain. First row: A chain of elliptical nanomagnets in the ground state with magnetization orientation indicated by arrows. The chain is in the ground state and hence the ordering is antiferromagnetic. Second row: Magnetization of the first magnet is flipped to change the input bit and the second magnet finds itself in a tied state where it experiences no net dipole interaction since the dipole interactions from its left and right neighbors cancel each other. Third row: The second and the third magnets are subjected to electrically induced stresses that rotate their magnetizations through a large angle while also depressing the shape anisotropy energy barriers within them. Fourth row: The second magnet is freed from stress and it switches to the desired "up" state since the dipole interaction from the left neighbor is now stronger than that from the right neighbor so that the tie is resolved. The right panel shows the energy landscape of the second magnet corresponding to the rows. Reproduced from *Appl. Phys. Lett.*, **97**, 173,105 (2010) with the permission of AIP Publishing

row. In that case, dipole coupling will enforce *anti-ferromagnetic ordering* in the ground state, meaning that nearest neighbors will have antiparallel magnetizations. This is shown in the top row of Fig. 6.5, where, of course, every magnetization is along the major axis of the ellipse since that is the easy axis. In this ground state arrangement, the bit encoded in the first nanomagnet on the left is repeated in every odd-numbered nanomagnet.

The second row in Fig. 6.5 shows the situation when the magnetization of the first nanomagnet is flipped to change the input bit. If one wants this new bit to ripple through the chain unidirectionally so that every odd numbered nanomagnet repeats the input bit, then every succeeding nanomagnet has to flip in a domino like fashion after the input bit is changed. That cannot happen for two reasons: (1) the dipole coupling between the two leftmost nanomagnets may not be strong enough to overcome the shape anisotropy energy barrier in the second nanomagnet and make its magnetization flip when the magnetization of the first nanomagnet is flipped. (2) More importantly, the effect of dipole interaction on the second nanomagnet due to its left and right neighbors are *exactly equal and opposite*, so that they cancel and the second nanomagnet feels no net dipole interaction effect at all. We call this a "tie state" since the influence from right and left neighbors are the same in a periodic chain of identical nanomagnets. In fact, because there is no net dipole interaction, the magnetization of the second nanomagnet will *not flip*, regardless of how small its internal energy barrier is.

The solution to this conundrum is shown in the third and fourth rows of Fig. 6.5. The second and third nanomagnets are first stressed to rotate their magnetizations by a large angle (up to 90°). The stress also depresses the energy barriers within them. Then the stress on the second magnetic is relaxed, that on the third is kept on, and the fourth is simultaneously stressed. When the second magnet is released from stress, it finds itself in an asymmetric environment since the magnetization orientations of its right and left neighbors are no longer antiparallel and hence the dipole interaction effect from them do not mutually cancel and the tie is broken. In fact, the influence from the left is now stronger. This will make the magnetization of the second magnet flip up (with greater than 50% probability). In the next cycle (not shown) the stress of the third will be released while the fourth and fifth will be placed under stress. That will make the third nanomagnet's magnetization flip down and the input bit would have rippled through the first three nanomagnets propagating unidirectionally from the left to right.

Reference [19] studied Bennett clocking in an array of Terfenol-D nanomagnets of major axis 102 nm, minor axis 98 nm and thickness 10 nm placed on a 40 nm thick PZT film for stress generation with a voltage. The simulations were carried out for 0 K temperature and did not consider thermal noise. It found that a bit can be propagated with 100% probability through one nanomagnet in 0.4 ns (clock rate 2.5 GHz) while dissipating 0.2 fJ of energy. If the clock rate is slowed to 1 GHz, then the dissipated energy can be reduced to less than 1 aJ. These studies were carried out for 0 K temperature and the figures would be obviously worse for room temperature operation.

6.2.5 Experimental Demonstration of Straintronic Bennett Clocking

Experimental demonstration of straintronic Bennett clocking was reported in Ref. [14]. In Fig. 6.5, one would need to generate *localized strain* to act on a pair of nanomagnets at a time. This would require placing gates around each nanomagnet, which would be a daunting fabrication challenge. Reference [14] sidestepped this challenge by using global strain which acts on all nanomagnets simultaneously. This would never enforce unidirectional propagation in a *uniform* chain of nanomagnets. One has to impose undirectionality either in *time*, as in Fig. 6.5, where pairs were being stressed sequentially and not simultaneously, or in *space,* by making each succeeding nanomagnet in the chain less elliptical than the preceding one so that the energy barrier decreased progressively down the chain. Reference [14] adopted the latter strategy. The nanomagnet shapes are shown in Fig. 6.6a.

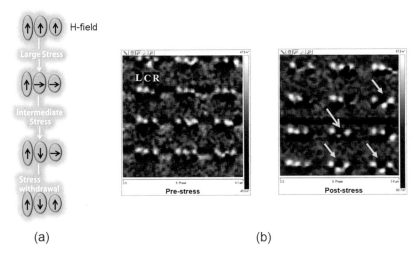

(a)   (b)

Fig. 6.6 Bennett clocking in a row of elliptical nanomagnets, with decreasing eccentricity, to propagate a bit encoded in the magnetization orientation unidirectionally from left to right along the chain using global strain. **a** Initially all magnetizations are aligned in the same direction with a magnetic field to make the ordering ferromagnetic. Then a high global stress is applied, ramped down to an intermediate value and ultimately relaxed to produce the anti-ferromagnetic ordering characteristic of Bennett clocking. **b** Magnetic force micrographs before and after the stress cycle showing that the anti-ferromagnetic ordering has been restored in four out of twelve 3-nanomagnet chains. Reproduced from *Nano Letters*, **16**, 1069 (2016) with permission of the American Chemical Society

All nanomagnets' magnetizations are initially aligned in the same direction with a strong magnetic field which enforces ferromagnetic ordering (nearest neighbors have parallel magnetizations), as shown in the top row of Fig. 6.6a. To achieve the anti-ferromagnetic ordering shown in the top row of Fig. 6.5, one proceeds as follows: At first a high global stress is applied which can invert the energy barriers in the two right nanomagnets in Fig. 6.6a and make their magnetizations align along the hard axis, but not in the far left nanomagnet which has the highest eccentricity and hence the largest energy barrier. Its magnetization remains aligned along its major axis. This is shown in the second row of Fig. 6.6a. Then the stress is reduced to an intermediate value which can keep the energy barrier in the far right nanomagnet inverted but not in the middle one. The magnetization of the former therefore remains aligned along its hard axis, but the magnetization of the latter rotates away from the hard axis and aligns along the easy axis. Because of unequal dipole interaction from the left and right neighbors, it aligns antiparallel to the magnetization of the left nanomagnet. This is shown in the third row of Fig. 6.6a. Finally, the global stress is withdrawn and now the far right nanomagnet also has to align along its easy axis (major axis), but because of the dipole influence of its left neighbor, it assumes an orientation that is antiparallel to that of the left neighbor. This is shown in the bottom row of Fig. 6.6. This, of course, restores that anti-ferromagnetic ordering which will ensure that the bit encoded in the first nanomagnet in the left is repeated in every odd-numbered nanomagnet and the input bit has propagated unidirectionally down the chain from left to right.

Figure 6.6b shows magnetic force micrographs of 12 trios of elliptical Co nanomagnets with decreasing eccentricity from left to right. The left panel shows the pre-stress condition where all magnetizations point in the same direction because a strong magnetic field has magnetized them in that direction. The right panel shows the states after the stress sequence: high, intermediate, relax. Four out of the 12 chains show restoration of the anti-ferromagnetic ordering and hence successful Bennett clocking. Again, the success rate is low, much lower than 50%, which once again confirms that dipole coupled straintronic logic is error-prone.

6.3 Switching Errors in Dipole Coupled Straintronic Boolean Logic Gates

A large number of studies on switching errors in dipole coupled straintronic gates (mostly inverters or Bennett clocked chains) have been reported [20–25]. They all considered *pristine* nanomagnets with no structural or material defects and the only source of error was thermal noise. Reference [20] studied an inverter (without any defect) and reported that the switching error probability at room temperature depends on the (1) dipole coupling strength, (2) stress levels, and (3) the stress ramp down rate.

Figure 6.7a shows the switching probability (probability of successful switching) in the inverter as a function of the center-to-center separation between the input and output nanomagnets (i.e. as a function of dipole coupling strength which varies as the inverse cube of the separation between neighbors) when the stress (applied selectively to the output nanomagnet only) is ramped down *abruptly*. The output nanomagnet has a major axis dimension of 105 nm, minor axis 95 nm and thickness 5.8 nm. The nanomagnet material is Terfenol-D. The plots are shown for various stress levels. Note that the successful switching probability progressively decreases as one increases the stress level because these are all super-critical stresses. Recall from the discussion accompanying Fig. 6.1 that the highest probability is obtained at the critical stress which just erodes (but does not invert) the shape anisotropy potential barrier within the nanomagnet. Exceeding this stress makes the switching probability progressively worse.

Figure 6.7b shows the switching probability as a function of the ramp down time when the separation between the nanomagnets is 200 nm. Note that the switching probability increases with increasing ramp down time. This feature is easy to understand. When the stress is ramped down from super-critical stress to zero stress, it must go through the critical stress where the switching probability is highest. The longer the system lingers around the critical stress, the higher the switching probability will be. As the ramp rate

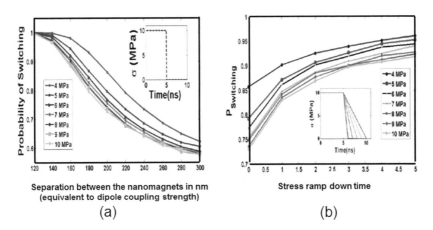

Fig. 6.7 **a** Switching success probability as a function of separation between the centers of the input and output nanomagnets along the line joining their centers in a dipole coupled straintronic inverter. The plots are shown for various stress magnitudes and the switching probability goes down with increasing stress. The stress is kept on for 5 ns and then ramped down abruptly. **b** The switching probability as a function of the ramp down time when the separation between the input and output nanomagnets is 200 nm. Reproduced *IEEE Trans. Nanotechnol.*, **12**, 1206, (2013) with permission of the Institute of Electrical and Electronics Engineers

becomes slower, the system spends more time at the critical stress level, which is why the probability of successful switching increases with increasing ramp down time.

Note that for any reasonable separation between the input and output nanomagnets, the switching error probability is still much higher than what is required for Boolean logic. Even under the best of circumstances, the error probability does not fall below 10^{-5} at room temperature. Therefore, Ref. [20] correctly concluded that dipole coupled straintronic logic is not viable.

Reference [23] showed that the error rate can be reduced by increasing the switching delay to about 10 ns and Ref. [24] attempted to reduce the error rate with creative pulse shaping of the strain pulse. The pulse shaping results are shown in Fig. 6.8. Pulse shaping helps, but the error rate is still as high as 10^{-3}, which makes straintronic dipole coupled logic ultimately impractical.

6.4   Switching Errors Caused by Defects and Imperfections

In the previous discussions of switching error in straintronic switching, the only source of error that was considered is thermal noise. A much more lethal source is structural defects which can cause the switching error probability to approach 100% at room temperature in some cases. This was studied in Ref. [26].

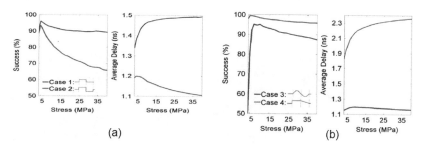

Fig. 6.8 **a** The room-temperature switching success probability as a function of stress in a strain-tronic dipole coupled inverter where the output nanomagnet is made of Terfenol-D and has a major axis of 105 nm, minor axis of 95 nm and thickness of 6 nm. The plots are shown for two different pulse shapes of the strain pulse. The switching delay is also shown as a function of stress. The switching success probability peaks at around 5 MPa which is slightly larger than the critical stress which was 4 MPa in this case. **b** The same plots for two other pulse shapes. Note that the probabilities are lower for Case 2 and Case 3 because in these two cases, the stress changes sign during one half-cycle and that actually increases the potential barrier within the output nanomagnet, which hinders switching. Reproduced from *Nanotechnology*, **26**, 245,202 (2015) with permission of the Institute of Physics

Figure 6.9 shows atomic force micrographs of real metallic nanomagnets fabricated by electron beam patterning of resists spun on a *piezoelectric* substrate, followed by development, metal evaporation and lift-off. The thickness of these nanomagnets are a few nm. Clearly, there are large scale thickness variations and other structural defects. Figure 6.9 shows four different types of defects that are labeled C2…C5. We mention in passing that ion beam milling may not a good fabrication option for these nanomagnets since the damage that will be done by the ion beam to the piezoelectric substrate could become intolerable.

For the purposes of micromagnetic simulations, carried out to calculate the straintronic switching probabilities of these "defective" nanomagnets under the influence of thermal noise, the above five types of defects were approximated as shown in Fig. 6.10.

The case that was examined with micromagnetic simulations dealt with the switching of a Terfenol-D elliptical nanomagnet of major axis 100 nm, minor axis 90 nm and thickness 6 nm. An isolated nanomagnet with in-plane magnetic anisotropy was considered. Its in-plane potential profile is a symmetric double well potential. Hence a small magnetic field of 3 mT was applied in one direction along the major axis to make the potential profile slightly asymmetric so that there is a preferred orientation for the magnetization, i.e. a non-degenerate ground state. This field could also be viewed as the effect of dipole coupling from a neighbor, had there been such a neighbor.

Let us assume that the initial magnetization is antiparallel to the preferred orientation as shown in Fig. 6.11a. The magnetic field alone cannot flip the magnetization to the ground state orientation since the energy barrier is too large to overcome as shown in Fig. 6.11b. Therefore uniaxial stress (compressive because Terfenol-D has positive magnetostriction)

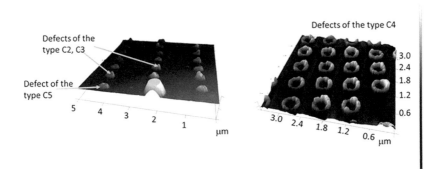

Fig. 6.9 Atomic force micrographs of four different types of structural defects denoted as C2…C5 formed when metallic nanomagnets of few nm thickness are delineated with electron beam patterning of resists, followed by development, metal evaporation and lift-off. Reprinted from D. Winters, M. A. Abeed, S. Sahoo, A. Barman and S. Bandyopadhyay, *Phys. Rev. Appl.*, **12**, 034,010 (2019) with permission of the American Physical Society. © 2019 American Physical Society

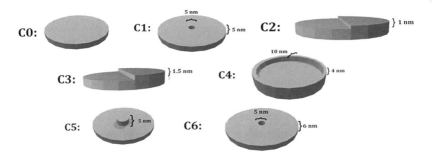

Fig. 6.10 The defects in Fig. 6.9 are approximated by these shapes for micromagnetic simulations of switching error probability in the presence of thermal noise at room temperature. C0 (no defect, an elliptical disk of major axis 100 nm, minor axis 90 nm, and thickness 6 nm), C1 (a shallow hole 5 nm in diameter and 5 nm deep at the center, not observed in Fig. 6.9, but still commonplace), C2 and C3 (one half of the nanomagnet thicker than the other by 1 and 1.5 nm, respectively), C4 (an annulus 10 nm thick and 4 nm high at the periphery; the height and thickness were kept uniform for ease of simulation. Introducing randomness in the height and thickness of the annulus will, if anything, exacerbate the error.), C5 (a raised cylindrical region 5 nm in diameter and 5 nm high), and C6 (a through hole 5 nm in diameter, not observed in Fig. 6.9). Reprinted from D. Winters, M. A. Abeed, S. Sahoo, A. Barman and S. Bandyopadhyay, *Phys. Rev. Appl.*, **12**, 034,010 (2019) with permission of the American Physical Society. © 2019 American Physical Society

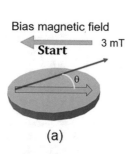
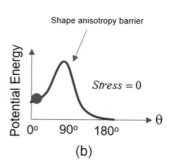

(a)   (b)

Fig. 6.11 a A defect-free Terfenol-D nanomagnet in the form of an elliptical disk of major axis 100 nm, minor axis 90 nm and thickness 6 nm. The magnetization initially points to the right along the major axis and a 3-mT magnetic field is applied in the opposite direction to make it flip to the left. **b** The potential energy profile of the nanomagnet in the presence of the magnetic field plotted as a function of the in-plane angle θ that the magnetization subtends with the initial direction. There is a local minimum at θ = 0° and a global minimum at θ = 180° due to the bias magnetic field. The magnetization is initially at the local minimum shown by the red ball. The shape anisotropy energy barrier prevents the magnetization from flipping to the left and reaching the global minimum. Stress will erode or invert this barrier to allow the magnetization to reach the global minimum and flip in the direction of the applied magnetic field. Reprinted from D. Winters, M. A. Abeed, S. Sahoo, A. Barman and S. Bandyopadhyay, *Phys. Rev. Appl.*, **12**, 034,010 (2019) with permission of the American Physical Society. © 2019 American Physical Society

has to be applied along the major axis to erode or invert the potential barrier and allow the magnetization to flip.

The switching scenario for the pristine defect-free nanomagnet is shown in Fig. 6.12. Let us say that we apply super-critical stress to invert the barrier. That will make the magnetization temporarily point along the minor axis (hard axis). Upon relaxation of stress, the magnetization will either go back to the initial orientation, which indicates a switching failure, or flip, which indicates a switching success. The magnetization dynamics in this case was simulated in the presence of room temperature thermal noise using micromagnetic simulations, and 1000 switching trajectories were generated. The switching error probability is the fraction of trajectories that failed to switch. These simulations were carried out for the pristine as well as defective nanomagnets, and the switching error probabilities as function of stress are shown in Fig. 6.13.

There are two important features to note in Fig. 6.13. First, there is a range of stress where the switching error probability is *lowest*, except for defect type C4. This is the neighborhood of the "critical stress" that was discussed earlier. Both sub-critical and super-critical stress increase the error probability. For the C4 type of defect, the critical stress is above 60 MPa. One cannot realistically apply more than 60 MPa to a Terfenol-D nanomagnet since the strain produced at that stress is 1200 ppm which may be close to the maximum strain that can be produced electrically in the piezoelectric [26]. Hence, one may not be able to reach the critical stress and attain the lowest switching error probability for the C4-type defect in the Terfenol-D nanomagnet.

The Defect Free Magnet

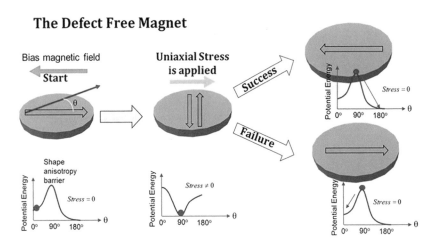

Fig. 6.12 Left panel: The initial magnetization state in a defect free nanomagnet. The magnetization is pointing to the right along the major (easy) axis and a magnetic field is applied in the opposite direction to make it flip. The magnetic field cannot flip the magnetization taking the system from the local to the global energy minimum because of the intervening shape anisotropy energy barrier. Middle panel: Sufficiently strong uniaxial stress applied along the major axis inverts the potential barrier and makes the magnetization point along the minor axis. Right panel: Upon stress release, the magnetization can either return to the initial state (which would be a switching failure) or flip (which would be a switching success). The error probability is the fraction of times the switching fails. Reprinted from D. Winters, M. A. Abeed, S. Sahoo, A. Barman and S. Bandyopadhyay, *Phys. Rev. Appl.*, **12**, 034,010 (2019) with permission of the American Physical Society. © 2019 American Physical Society

Second the error probabilities are very sensitive to the type of defect. Generally speaking, extended defects like C3 and C4 are much more lethal than localized defects like C1 and C6. Because extended defects may be unavoidable in a large ensemble of nanomagnets, Ref. [26] questioned the viability of straintronic "Boolean logic" which has extremely stringent reliability requirements (error probability $< 10^{-15}$). Reference [26] also opined that because of this, the niche for straintronics may be in unconventional non-Boolean computing like neuromorphic, probabilistic, or Shannon inspired statistical computing [27].

6.5 Relatively Error-Resilient Straintronic Universal Logic Gate not Based on Dipole Coupling

There are other versions of straintronic logic, *not* based on dipole coupling, which are much more reliable, although probably still not sufficiently reliable for Boolean logic. One of them was proposed in Ref. [2]. It is shown schematically in Fig. 6.14. At the

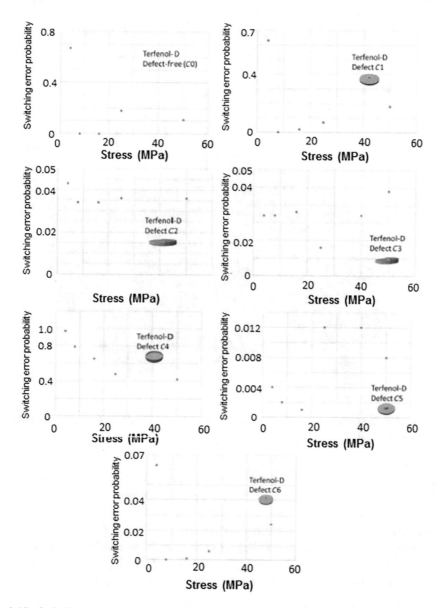

Fig. 6.13 Switching error probability versus stress at room temperature for the magnetic field of 3 mT. Reproduced from D. Winters, M. A. Abeed, S. Sahoo, A. Barman and S. Bandyopadhyay, *Phys. Rev. Appl.*, **12**, 034,010 (2019) with permission of the American Physical Society. © 2019 American Physical Society

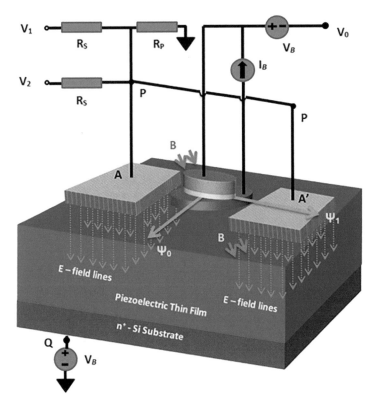

Fig. 6.14 A straintronic error-resilient NAND gate implemented with a skewed straintronic magnetic tunnel junction

heart of the gate is a skewed straintronic magnetic tunnel junction where the easy axes of the hard and soft layers are non-collinear. An in-plane magnetic field is applied along the hard axis of the soft layer to bring the two stable magnetization orientations out of the latter's easy axis and locate them perpendicular to each other in the manner of Fig. 4.5. The calculated room temperature switching error probability was 10^{-8} and the energy-delay product was 1.6×10^{-26} J-s [2]. The operation of this logic gate is complex and hence not described here, but the interested reader can refer to ref. [2].

References

1. B. Behin-Aein, D. Datta, S. Salahuddin, S. Datta, Proposal for an all-spin logic device with built-in memory. Nat. Nanotechnol. **5**, 266 (2010)
2. A.K. Biswas, J. Atulasimha, S. Bandyopadhyay, An error-resilient non-volatile magneto-elastic universal logic gate with ultralow energy delay product. Sci. Rep. **4**, 7553 (2014)
3. R. Waser, *Nanoelectronics and Information Technology* (Wiley, NJ, USA, 2003). (Chapter 3)
4. S. Bandyopadhyay, Nanomagnetic Boolean logic: The tempered (and realistic) vision. IEEE Access **9**, 7743 (2021)
5. R.P. Cowburn, M.E. Welland, Room temperature magnetic quantum cellular automata. Science **287**, 1466 (2000)
6. S. Bandyopadhyay, B. Das, A.E. Miller, Supercomputing with spin polarized single electrons in a quantum coupled architecture. Nanotechnology **5**, 113 (1994)
7. H. Agarwal, S. Pramanik, S. Bandyopadhyay, Single spin universal Boolean logic gate. New. J. Phys. **10**, 015001 (2008)
8. A. Imre et al., Majority logic gate for magnetic quantum-dot cellular automata. Science **311**, 205 (2006)
9. S. Bandyopadhyay, M. Cahay, Electron spin for classical information processing: A brief survey of spin-based logic devices, gates and circuits. Nanotechnology **20**, 412001 (2009)
10. S. Breitkreutz, J. Kiermaier, I. Eichwald, X. Ju, G. Csaba, D. Schmitt-Landsiedel, M. Becherer, Majority gate for nanomagnetic logic with perpendicular magnetic anisotropy. IEEE Trans. Magn. **48**, 4336 (2012)
11. I. Eichwald, A. Bartel, J. Kiermaier, S. Breitkruetz, G. Csaba, D. Schmitt-Landsiedel, M. Becherer, Nanomagnetic logic: Errorfree, directed signal transmission by an inverter chain. IEEE Trans. Magn. **48**, 4332 (2012)
12. J. Von Neumann, Probabilistic logics and the synthesis of reliable organisms from unreliable components. Autom. Stud. **34**, 43–98 (1956)
13. A.D. Patil, S. Manipatruni, D. Nikonov, I.A. Young, N.R. Shanbhag, Shannon-inspired statistical computing to enable spintronics (2017). arXiv:1702.06119
14. N. D'Souza, M. Salehi-Fashami, S. Bandyopadhyay, J. Atulasimha, Experimental clocking of nanomagnets with strain for ultra low power Boolean logic. Nano Lett. **16**, 1069 (2016)
15. H. Ahmad, J. Atulasimha, S. Bandyopadhyay, Electric field control of magnetic states in isolated and dipole-coupled FeGa nanomagnets delineated on a PMN-PT substrate. Nanotechnology **26**, 401001 (2015)
16. M. Salehi-Fashami, J. Atulasimha, S. Bandyopadhyay, Magnetization dynamics, throughput and energy dissipation in a universal multiferroic nanomagnetic logic gate with fan-in and fan-out. Nanotechnology **23**, 105201 (2012)
17. C.H. Bennett, The thermodynamics of computation: A review. Int. J. Theor. Phys. **21**, 905 (1982)
18. J. Atulasimha, S. Bandyopadhyay, Bennett clocking of nanomagnetic logic using multiferroic single-domain nanomagnets. Appl. Phys. Lett. **97**, 173105 (2010)
19. M. Salehi-Fashami, K. Roy, J. Atulasimha, S. Bandyopadhyay, Magnetization dynamics, Bennett clocking and associated energy dissipation in multiferroic logic. Nanotechnology **22**, 155201 (2011)
20. M.S. Fashami, K. Munira, S. Bandyopadhyay, A.W. Ghosh, J. Atulasimha, Switching of dipole coupled multiferroic nanomagnets in the presence of thermal noise: Reliability of nanomagnetic logic. IEEE Trans. Nanotechnol. **12**, 1206 (2013)

21. M.M. Al-Rashid, D. Bhattacharya, S. Bandyopadhyay, J. Atulasimha, Effect of nanomagnet geometry on reliability, energy dissipation, and clock speed in strain-clocked DC-NML. IEEE Trans. Electron. Dev. **62**, 2978 (2015)
22. K. Munira, Y. Xie, S. Nadri, M.B. Forgues, M.S. Fashami, J. Atulasimha, S. Bandyopadhyay, A.W. Ghosh, Reducing error rates in straintronic multiferroic nanomagnetic logic by pulse shaping. Nanotechnology **26**, 245202 (2015)
23. M.M. Al-Rashid, S. Bandyopadhyay, J. Atulasimha, Dynamic error in strain-induced magnetization reversal of nanomagnets due to incoherent switching and formation of metastable states: A size-dependent study. IEEE Trans. Electron Devices **63**, 3307 (2016)
24. F.M. Spedalieri, A.P. Jacob, D.E. Nikonov, V.P. Roychowdhury, Performance of magnetic quantum cellular automata and limitations due to thermal noise. IEEE Trans. Nanotechnol. **10**, 537 (2011)
25. D. Carlton, B. Lambson, A. Scholl, A. Young, P. Ashby, S. Dhuey, J. Bokor, Investigation of defects and errors in nanomagnetic logic circuits. IEEE Trans. Nanotechnol. **11**, 760 (2012)
26. D. Winters, M.A. Abeed, S. Sahoo, A. Barman, S. Bandyopadhyay, Reliability of magnetoelastic switching of non-ideal nanomagnets with defects: A case study for the viability of straintronic logic and memory. Phys. Rev. Appl. **12**, 034010 (2019)
27. N.R. Shanbhag, N. Verma, Y. Kim, A.D. Patil, L.R. Varshney, Shannon-inspired statistical computing for the nanoscale era. Proc. IEEE **107**, 90 (2019)

Switching the Magnetizations of Magnetostrictive Nanomagnets with Time Varying Periodic Strain (Surface Acoustic Waves)

7

The use of surface acoustic waves (SAW) to switch the magnetization of magnetostrictive nanomagnets using the inverse magnetostriction (Villari) effect—also sometimes referred to as the magneto-elastic effect—has a long history. A SAW exerts (periodic) time-varying strain (instead of static strain) on a magnetostrictive nanomagnet and hence should induce time-varying periodic rotation of the magnetization via the Villari effect. Reference [1] reported periodic switching of magnetization between the hard and the easy axes of 40 μm $\times$ 10 μm $\times$ 10 nm Co bars sputtered on a piezoelectric LiNbO$_3$ film in which a SAW was launched. Acoustically induced switching has also been observed in thin films [2] and later extended to focusing surface acoustic waves (SAW) to switch a specific spot in an iron-gallium film [3]. The frequency and wave vector of the SAW can affect the switching [4]. It has been claimed that it is possible to flip magnetization (180° rotation) with an appropriately timed acoustic pulse [5]. In this context, stroboscopic X-ray techniques were employed to probe strain waves and their influence on the magnetization of magnetostrictive magnets at the nanoscale [6]. SAW has also been proposed as a mechanism for depinning domain walls and drive domain wall motion in magnetostrictive magnets [7], although actually moving domain walls with the velocity of a SAW, which would be a remarkable advance, is still elusive at the time of this writing. At present domain wall velocity is much smaller than SAW velocity in most piezoelectric crystals.

A surface acoustic wave consists of acoustic phonons in the piezoelectric material. A very important field of current research is phonon-magnon coupling which essentially refers to excitation of spin waves in two-phase multiferroics (a magnetostrictive magnet elastically coupled to an underlying piezoelectric film or substrate) by acoustic waves. We will discuss this field in a later chapter in more detail. Spin waves were excited in GaMnAs layers (weakly magnetostrictive dilute magnetic semiconductor) by a picosecond strain pulse [8] which could also initiate magnetization dynamics in GaMnAs [9] and GaMn(As, P) [10]. In nanomagnets with in-plane anisotropy, surface acoustic waves have been utilized to drive ferromagnetic resonance in thin Ni films [11, 12]. Resonant effects

S. Bandyopadhyay, *Magnetic Straintronics*, Synthesis Lectures on Engineering, Science, and Technology, https://doi.org/10.1007/978-3-031-20683-2_7

were studied by spatial mapping of focused SAWs [13]. Theoretical studies have revealed the possibility of complete magnetization reversal in a nanomagnet subjected to acoustic wave pulses [14].

A recent perspective on spintronics with surface acoustic devices can be found in Ref. [15].

7.1   Switching an Isolated Magnetostrictive Nanomagnet with Time-Varying Strain (Acoustic Wave)

One of the earliest experiments of switching a magnetostrictive nanomagnet with an acoustic wave showed that an isolated nanomagnet, fabricated on a piezoelectric substrate, can undergo a *permanent* change in its magnetic state (i.e. the magnetic ordering within the nanomagnet) when subjected to a radio frequency (RF) acoustic wave (AW) of sufficient amplitude. The nanomagnet does not revert back to its original state once the AW is withdrawn, showing that the change in the magnetic state is irreversible. This phenomenon was demonstrated experimentally in elliptical cobalt nanomagnets of major axis dimension 340 nm, minor axis dimension 270 nm and thickness 12 nm [16]. The experimental observation of this irreversible transition was replicated by micromagnetic simulations of magneto-dynamics driven by magneto-elastic coupling of the magnetization to the AW.

The magnetostrictive nanomagnets were fabricated on a piezoelectric substrate between interdigitated transducers (IDTs) and the inter-magnet spacing was large enough that there was negligible dipole interaction between neighbors. In other words, they acted as *isolated* nanomagnets. First, they were all magnetized in the same direction as shown in Fig. 7.1a with a 0.2 T magnetic field. An AW of 3.5 MHz frequency was then launched in the substrate by applying a peak-to-peak voltage of 3.5 V and 3.5 MHz frequency between the adjacent fingers of IDTs, as shown in Fig. 7.1b. The wavelength of this wave was comparable to the surface thickness, which is why it was not confined to the surface of the substrate and penetrated into the bulk. Hence, it is called an acoustic wave rather than a surface acoustic wave.

The magnetic ordering in the nanomagnets were ascertained with magnetic force micrograph before launching the AW and showed that each magnet was magnetized in the direction of the magnetic field. The AW was subsequently launched and drove the magnetization in each nanomagnet into a vortex state, which persisted after the SAW was withdrawn. Only a magnetic field could restore the original magnetic state, as shown in Fig. 7.1c.

An acoustic wave will alternately apply tensile and compressive stress on the nanomagnets. Micromagnetic simulations were carried out to examine the effect of such stress variations on the magnetic state of an isolated nanomagnet and revealed that both signs of stress will take the magnetization from a nearly single domain state to a vortex state and the nanomagnet will remain in that state (non-volatility) after stress is withdrawn. This is shown in Fig. 7.2.

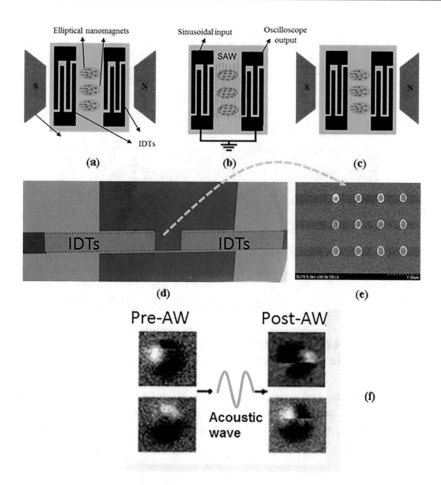

Fig. 7.1 a Cobalt nanomagnets are fabricated on a piezoelectric substrate between interdigitated transducers (IDTs) and magnetized into single domain states (all spins pointing in the same direction) with a magnetic field. **b** The magnetic field is withdrawn and an acoustic wave (AW) is launched with one set of IDTs, which drives the nanomagnet out of the single domain state into a vortex state. The state persists after the AW is withdrawn. **c** Application of a magnetic field restores the original state. **d** Scanning electron micrograph showing the IDTs. The red square contains the nanomagnets. **e** Magnified view of the red square showing the nanomagnets within. **f** Magnetic force micrograph showing that the AW takes the magnetic ordering from a single domain state to a vortex state. Reproduced from *Nano Letters*, **16**, 5681 (2016) with permission of the American Chemical Society

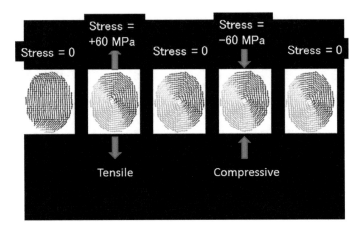

Fig. 7.2 Results of micromagnetic simulations of spin texture within a single nanomagnet under a cycle of tensile stress, followed by relaxation, compressive stress and relaxation. For the amplitude of the AW that was launched, the stress amplitude was calculated to be 60 MPa. Reproduced from *Nano Letters*, **16**, 5681 (2016) with permission of the American Chemical Society

7.2 A Dipole-Coupled Straintronic Inverter Clocked with an Acoustic Wave

A straintronic inverter, fashioned out of two elliptical magnetostrictive nanomagnets placed close to each other in order to have dipole coupling (recall the description in the previous chapter), was clocked with an AW instead of static strain [17] to test if it performed as a NOT gate. The AW generates time varying strain and that triggered the inverter operation, just like static strain did (recall the previous chapter).

The nanomagnet pair making up the inverter is shown in Fig. 7.3a. As usual, the nanomagnet hosting the input bit had a higher degree of eccentricity than the nanomagnet hosting the output bit. This made the shape anisotropy energy barrier in the input nanomagnet much higher than that in the output magnet, so that the periodic strain generated by the AW did not affect the magnetization of the input nanomagnet but could affect the magnetization of the output nanomagnet (again recall the description in the previous chapter).

In this case, the SAW did not drive the magnetization permanently into the vortex states because of the presence of dipole coupling (these are not isolated nanomagnets, but interacting nanomagnets). The dipole coupling makes a difference.

The inverter operation was demonstrated in the same way as described in the previous chapter. A magnetic field aligned the two magnetizations parallel to each other, representing the bit state (1, 1) where both input and output binary bits are 1. This state was verified with magnetic force microscope. A 5-MHz AW was then launched and, after many cycles, stopped. The maximum strain generated by the wave was estimated to be

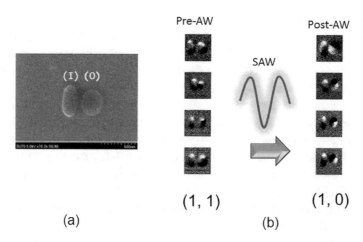

Pre-AW   Post-AW

SAW

(1, 1)   (1, 0)

(a)   (b)

Fig. 7.3 **a** Scanning electron micrographs of closely spaced dipole coupled elliptical nanomagnets, one more eccentric than the other, forming an inverter. The more eccentric member hosts the input bit (I) and the other the output bit (O). **b** Magnetic force micrographs of four different inverters showing successful Boolean inversion taking place in all four. Reproduced from *Appl. Phys. Lett.*, **109**, 102,403 (2016) with permission of AIP Publishing

142 ppm [17]. Examination with magnetic force microscope of four different inverter pairs showed that the inverter operation took place successfully in all four and the final bit states became (1, 0) where the output bit is the correct logic complement of the input bit [17]. This is shown in Fig. 7.3b.

Curiously, gate error probability in this case is *much lower* that what one observes when the inverter is clocked with *static* strain (recall Chap. 6). Since four out of four pairs switched, we can say with some certainty that the error probability is smaller than what one obtains with static strain. Why this is the case is not well understood, but it is conjectured that time varying strain *repeatedly* stresses the nanomagnets, cycle after cycle, subjecting it to alternating cycles of tensile and compressive strain, and this increases the probability that they will ultimately switch.

7.3   Switching a Magnetic Tunnel Junction with a Mixture of Spin Transfer Torque and *Resonant* Surface Acoustic Waves

We had mentioned earlier (in Sect. 5.2 in Chap. 5) the idea of using mixed mode switching (STT+SAW) to reduce the switching energy dissipation in a magnetic tunnel junction (MTJ) switched with STT, such as STT-RAM. This reduction becomes much larger with the use of *resonant* SAW in a p-MTJ where the soft layer has perpendicular magnetic anisotropy. This is an important scenario for two reasons. First, we are dealing with perpendicular magnetic anisotropy (instead of in-plane) which lends itself to smaller MTJ cross-sections, and is therefore preferable for memory applications. Second, the reduction in the STT switching current (and hence the switching energy dissipation) is much larger.

Resonant SAW implies that the SAW frequency is tuned to the ferromagnetic resonance frequency of the soft layer. This can make the magnetization of the soft layer deflect by very large angles from the easy axis, even at relatively low SAW power, because the deflection resonantly builds up over a few cycles. In fact, the SAW alone makes the magnetization process in a cone with a large deflection of ∼45° from the easy axis, thereby drastically reducing the STT current required to flip the magnetization, while keeping the STT duration and the switching error probability the same [18].

Figure 7.4 illustrates the idea. The resonant SAW is a periodic strain that acts like a periodic effective magnetic field at the ferromagnetic resonance (FMR) frequency. It precesses the magnetization through large angles since the energy from the SAW is (magneto-elastically) coupled into the magnetization precession over many cycles. Figure 7.5a shows the manner in which such resonant SAW drives the magnetization of a perpendicular magnetic tunnel junction (p-MTJ) to precess in a cone with large deflection from the easy axis of magnetization [18]. This reduces the STT current needed to flip the magnetization much more than the case where the STT switching is not assisted by *resonant* SAW as shown in Fig. 7.5b.

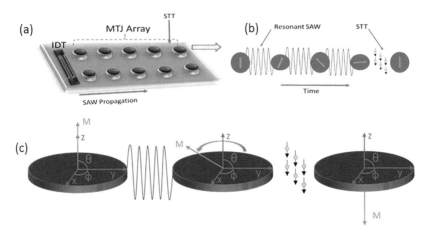

Fig. 7.4 **a** A magnetic tunnel junction (MTJ) array switched with resonant surface acoustic wave (SAW) and spin-transfer-torque (STT). The array is fabricated on a piezoelectric substrate and a SAW is launched into the substrate by applying an oscillating voltage at the interdigitated transducer (IDT). **b** Magnetization dynamics (time evolution of the magnetization) in the soft layer of the MTJ with resonant SAW+STT switching of in-plane magnetization in an MTJ with in-plane magnetic anisotropy and **c** Magnetization dynamics with resonant SAW+STT switching of out-of-plane magnetization in an MTJ with perpendicular magnetic anisotropy. Reproduced from *Appl. Phys. Lett.*, **115**, 112,405 (2019) with the permission of AIP Publishing

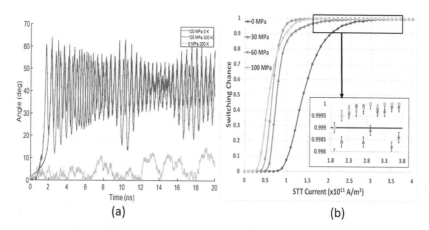

Fig. 7.5 a Out-of-plane magnetization dynamics simulations showing the time evolution of the magnetization precession angle under resonant SAW generating sinusoidal stress of amplitude 100 MPa with and without thermal noise and under purely thermal noise (no stress). **b** Switching probability vs. STT current density for three different SAW amplitudes as well as for no SAW applied. Reproduced from *Appl. Phys. Lett.*, **115**, 112,405 (2019) with the permission of AIP Publishing

7.4   Simulated Annealing in a Two-Dimensional Periodic Array of Magnetostrictive Nanomagnets Actuated by a Surface Acoustic Wave

Switching the magnetization of a magnetostrictive nanomagnet with a SAW has many applications that go well beyond inverters, Boolean logic or memory. In fact, they may have more important applications in *non-Boolean* computing paradigms which are beginning to capture the limelight because of superior energy efficiency, ability to solve intractable (NP-hard or NP-complete) problems more easily and better error-resilience. It turns out that magnetic or multiferroic platforms may be very synergistic with non-Boolean computing and provide excellent hardware accelerators for performing specific tasks in machine learning and artificial intelligence. A popular (non-Boolean) computing approach for solving difficult combinatorial optimization problems (e.g., max-cut, min-cut, or traveling salesperson) is a *Boltzmann machine*. This is a network of interacting bistable devices (referred to as "binary neurons" in neuromorphic computing literature) where the interaction is viewed as an interconnection. In the neuromorphic computing literature, these interconnections are called "synapses". The interaction strength (and hence the weight of the synapse) should ideally be variable individually.

In a Boltzmann machine, a problem is hardware-mapped on to an array of interconnected binary neurons (the binary neurons could be bistable nanomagnets with their two stable magnetization orientations encoding the binary bits 0 and 1, and the interconnections could be nearest neighbor dipole coupling). The mapping is carried out by biasing the neurons in a certain way and adjusting the interconnection strengths, so that when the interconnected system is allowed to relax to the ground state, the bit states in the neurons will represent the optimal solution to a problem. One first defines a "cost-function" for the system given by $E = [S]^T[x] + [S]^T[w][S]$ where $[S]$ is $n \times 1$ column vector whose elements are 0 or 1 (representing the states of the binary neurons or bistable nanomagnets), $[x]$ is a bias vector which is also a $n \times 1$ column and $[w]$ is a symmetric $n \times n$ matrix called the "weight matrix". An element of the $[w]$ matrix W_{ij} represents the interconnection strength between the i-th and the j-th binary neuron, i.e. the effectiveness of dipole coupling between the i-th and the j-th nanomagnet. To solve a given problem, the bias vector $[x]$ and the weight matrix $[w]$ are picked such that the value of $[S]$ (the states of the neurons) which will minimize the cost function E represents the optimal solution. The cost function conforms to the energy. If the array is left to itself, it will decay to the minimum energy state or ground state according to the laws of thermodynamics (minimizing the cost function) and thereafter if we probe the states of the neurons (bistable nanomagnets) to find the value of $[S]$ in the ground state, we will find the solution. The ground state value of $[S]$ represents the optimal solution. The reader will understand that in this case, we do not know the value of $[S]$ corresponding to the optimal solution apriori, but we do know the value of $[w]$ and $[x]$ which will yield the optimal solution and these we adjust to obtain the optimal solution through the ground state value of $[S]$. This is a hardware approach to solving an optimization problem and is much faster than any software based approach since we do not have to execute instruction sets sequentially, which is what we must do in software.

The reason these constructs are called "Boltzmann machines" is that the probability of relaxing to the i-th energy state is given $p_i = e^{-\beta E_i} / \sum_i e^{-\beta E_i}$ (β is a constant) in accordance with Boltzmann statistics. Hence, relaxing to the ground state, which is the lowest energy state, has the highest probability. Thus, the optimal solution is found with the highest probability. Unfortunately, that highest probability is not unity (it is less than 1) and hence there is always some likelihood that the optimal solution will not be found. In that case, one performs a procedure called "simulated annealing" to find the optimal solution. More on this later.

Boltzmann machines are often built with interacting nanomagnetic systems [19–23] because they are particularly suitable for this approach. Dipole interaction acts like the synapse and although it is very difficult to vary its weight arbitrarily, the fact that this "synapse" is wireless, dissipates no energy and consumes no area on a chip, makes the idea of implementing Boltzmann machines with dipole coupled bistable nanomagnets attractive. There are some instances where uniform synapse weight can provide useful computation and in those cases these magnetic implementations become very attractive.

The Achilles' heel of ground state computing, however, is the possibility that the Boltzmann machine, namely the magnetic array, may get stuck in a metastable state and fail to relax to the ground state. The magnetic states of the nanomagnets (0 or 1) in the metastable state do not represent the optimal solution; instead, they represent a suboptimal solution. If this happens, a procedure called *simulated annealing* can drive the system out of the metastable state and take it to the ground state to produce the optimal solution.

Simulated annealing of this type was mimicked in an interacting nanomagnetic system comprising a two-dimensional array of 3 × 3 dipole-coupled elliptical magnetostrictive nanomagnets deposited on a piezoelectric substrate and excited by a SAW [24]. In the *ground state* of the array, the magnetizations along any row would be anti-ferromagnetically ordered, while along any column, it will be ferromagnetically ordered, as shown in Fig. 7.6, because of dipole coupling between neighbors, which prefers this arrangement. Here, the dipole coupling strength between any two pairs in every row is the same and similarly the dipole coupling strength between any two pairs in every column in the same. The system was intentionally driven out of the ground state with an external agent and thereafter pinned in a metastable state where the magnetic ordering was different from the ground state ordering. This was a metastable state and thermal perturbations could not make the array escape into the ground state. However, when a surface acoustic wave was launched in the substrate, it released the array from the metastable state and returned it to the ground state, thus implementing simulated annealing.

Figure 7.7a shows the atomic force micrograph of a 3 × 3 array of Co nanomagnets (selected from a much larger array) fabricated on a $LiNbO_3$ substrate. Two pairs of gold contact pads were delineated on the substrate, each pair on one side of the nanomagnet assembly, to launch surface acoustic wave (SAW) in the substrate.

The MFM image of the selected 3 × 3 array is shown in Fig. 7.8a. One can clearly see the ordering depicted in Fig. 7.6, where the nanomagnets along a column are magnetized in the same direction and alternating columns have opposite directions of magnetization, i.e. the system is in the ground state. This image was obtained with a low-moment MFM tip in order to carry out non-invasive imaging. Next, the array was perturbed with a high-moment MFM tip to drive it out of the ground state and the MFM image after

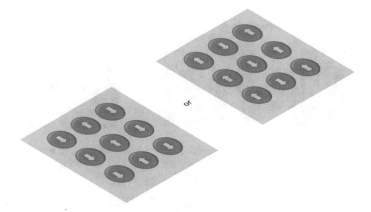

Fig. 7.6 Ground state magnetization orientations in a 3 × 3 array of dipole coupled elliptical nano-magnets with in-plane magnetic anisotropy. The ordering of the magnetizations is ferromagnetic along columns and anti-ferromagnetic along rows in the ground state. Dipole coupling between neighbors enforces this ground state configuration. Adapted from *J. Phys. D: Appl. Phys.*, **53**, 445,002 (2020) with permission of the Institute of Physics

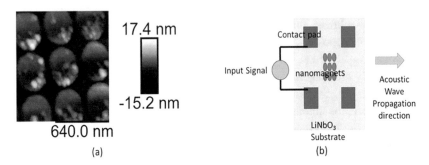

Fig. 7.7 **a** Atomic force micrograph of a 3×3 array of dipole coupled elliptical cobalt nanomagnets with in-plane anisotropy fabricated on a LiNbO₃ substrate. **b** The nanomagnets are flanked by contact pads that are used to launch a surface acoustic wave (SAW) in the substrate. The SAW performs the action of simulated annealing. Reproduced from *J. Phys. D: Appl. Phys.*, **53**, 445,002 (2020) with permission of the Institute of Physics

the perturbation can be seen in Fig. 7.8b. Clearly, the ground state ordering has been destroyed and the system has not spontaneously returned to the ground state (i.e. it is stuck in a metastable state). c. After subjecting the array to a SAW, the ground state configuration is restored. Hence the SAW drove the system out of the metastable state and into the ground state, thereby performing the act of simulated annealing.

After launching a surface acoustic wave and waiting for a few seconds, the nanomagnets were imaged again. Their magnetizations were found to have returned to the ground state, indicating successful "simulated annealing" [24]. The SAW acted as the annealing agent in this case.

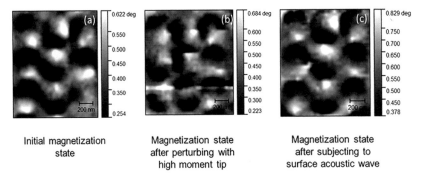

Initial magnetization state

Magnetization state after perturbing with high moment tip

Magnetization state after subjecting to surface acoustic wave

Fig. 7.8 Magnetic force microscopy (MFM) images of: **a** The initial magnetic ordering of a 3×3 array of elliptical Co nanomagnets, which corresponds to the ground state. **b** The ordering of the array after perturbation with a high moment MFM tip (this is a metastable state). **c** The ordering after the passage of the surface acoustic wave. The system gets unpinned from the metastable state and returns to the ground state. Reproduced from *J. Phys. D: Appl. Phys.*, **53**, 445,002 (2020) with permission of the Institute of Physics

These applications of straintronics to non-Boolean computing are relatively error-resilient since here the interactions between multiple neurons (nanomagnets) elicits the computational activity. We will examine many error-resilient non-Boolean computing applications enabled by straintronics in Chap. 10.

References

1. S. Davis, A. Baruth, S. Adenwalla, Magnetization dynamics triggered by surface acoustic waves. Appl. Phys. Lett. **97**, 232507 (2010)
2. W. Li, B. Buford, A. Jander, P. Dhagat, Acoustically assisted *magnetic* recording: A new paradigm in *magnetic* data storage. IEEE Trans. Magn. **50**, 37–40 (2014)
3. W. Li, B. Buford, A. Jander, P. Dhagat, Writing magnetic patterns with surface acoustic waves. J. Appl. Phys. **115**, 17E307 (2014)
4. P. Kuszewski et al., Resonant magneto-acoustic switching: influence of Rayleigh wave frequency and wave vector. J. Phys. Condens. Matter **30**, 244003 (2018)
5. O. Kovalenko, T. Pezeril, V.V. Temnov, New concept for magnetization switching by ultrafast acoustic pulses. Phys. Rev. Lett. **110**, 266602 (2013)
6. M. Foerster, F. Macià, N. Statuto, S. Finizio, A. Hernández-Mínguez, S. Lendínez, P.V. Santos, J. Fontcuberta, J.M. Hernàndez, M. Kläui, L. Aballe, Direct imaging of delayed magneto-dynamic modes induced by surface acoustic waves. Nat. Commun. **8**, 407 (2017)
7. A. Adhikari, E. Gilroy, T. Hayward, S. Adenwalla, Surface acoustic wave assisted depinning of magnetic domain walls. J. Phys. D: Appl. Phys. **33**, 31LT01 (2021)
8. M. Bombeck, A.S. Salasyuk, B.A. Glavin, A.V. Scherbakov, C. Brüggemann, D.R. Yakovlev, V.F. Sapega, X. Liu, J.K. Furdyna, S.V. Akimov, M. Bayer, Excitation of spin waves in ferromagnetic (Ga, Mn)As layers by picosecond strain pulses. Phys. Rev. B **85**, 195324 (2012)
9. A.V. Scherbakov, A.S. Salasyuk, A.V. Akimov, X. Liu, M. Bombeck, C. Brüggemann, D.R. Yakovlev, V.F. Sapega, J.K. Furdyna, M. Bayer, Coherent magnetization precession in ferromagnetic (Ga, Mn)As induced by picosecond acoustic pulses. Phys. Rev. Lett. **105**, 117204 (2010)
10. L. Thevenard, E. Peronne, C. Gourdon, C. Testelin, M. Cubukcu, E. Charron, S. Vincent, A. Lemaître, B. Perrin, Effect of picosecond strain pulses on thin layers of the ferromagnetic semiconductor (Ga, Mn)(As, P). Phys. Rev. B **82**, 104422 (2010)
11. M. Weiler, L. Dreher, C. Heeg, H. Huebl, R. Gross, M.S. Brandt, S.T.B. Goennenwein, Elastically driven ferromagnetic resonance in nickel thin films. Phys. Rev. Lett. **106**, 117601 (2011)
12. J. Janušonis, C.L. Chang, P.H.M. Loosdrecht, R.I. Tobey, Frequency tunable surface magneto elastic waves. Appl. Phys. Lett. **106**, 181601 (2015)
13. U. Singh, S. Adenwalla, Spatial mapping of focused surface acoustic waves in the investigation of high frequency strain induced changes. Nanotechnology **26**, 255707 (2015)
14. L. Thevenard, J.Y. Duquesne, E. Peronne, H.J. von Bardeleben, H. Jaffres, S. Ruttala, J.M. George, A. Lemaître, C. Gourdon, Irreversible magnetization switching using surface acoustic waves. Phys. Rev. B **87**, 144402 (2013)
15. J. Puebla, Y. Hwang, S. Maekawa, Y. Otani, Perspectives on spintronics with surface acoustic waves. Appl. Phys. Lett. **120**, 220502 (2022)
16. V. Sampath, N. D'Souza, D. Bhattacharya, G.M. Atkinson, S. Bandyopadhyay, J. Atulasimha, Acoustic wave induced magnetization switching of magnetostrictive nanomagnets from single domain to non-volatile vortex states. Nano Lett. **16**, 5681 (2016)

17. V. Sampath, N. D'Souza, G.M. Atkinson, S. Bandyopadhyay, J. Atulasimha, Experimental demonstration of acoustic wave induced magnetization switching in dipole coupled magnetostrictive nanomagnets for ultralow power computing. Appl. Phys. Lett. **109**, 102403 (2016)
18. A. Roe, D. Bhattacharya, J. Atulasimha, Resonant acoustic wave assisted spin-transfer-torque switching of nanomagnets. Appl. Phys. Lett. **115**, 112405 (2019)
19. N. D'Souza, J. Atulasimha, S. Bandyopadhyay, An ultrafast image recovery and recognition system implemented with nanomagnets possessing biaxial magneto-crystalline anisotropy. IEEE Trans. Nanotechnol. **11**, 896 (2012)
20. A. Payatakov, A. Kaminskiy, E. Lomov, W. Ren, S. Cao, A. Zvezdin, Routes to low energy magnetic electronics. SPIN **9**, 1940004 (2019)
21. B. Sutton, K.Y. Camsari, B. Behin-Aein, S. Datta, Intrinsic optimization using stochastic nanomagnets. Sci. Rep. **7**, 44370 (2017)
22. S. Nasrin, J.L. Drobitch, S. Bandyopadhyay, A.R. Trivedi, Low-power restricted Boltzmann machine using mixed mode magneto-tunneling junctions. IEEE Elec. Dev. Lett. **40**, 345 (2019)
23. S. Bhanja, D.K. Karunaratne, R. Panchumarthy, S. Rajaram, S. Sarkar, Non-Boolean computing with nanomagnets for computer vision applications. Nat. Nanotechnol. **11**, 177 (2016)
24. M.A. Abeed, S. Bandyopadhyay, Simulated annealing with surface acoustic wave in a dipole-coupled array of magnetostrictive nanomagnets for collective ground state computing. J. Phys. D: Appl. Phys. **53**, 445002 (2020)

Analog Straintronics

8

Most of the discussions in the previous chapters had focused on digital applications where straintronics is utilized to switch a bistable magnetostrictive nanomagnet from one state encoding one binary bit to the other encoding the other binary bit. However, there are analog applications of magnetic straintronics that are equally intriguing and useful. This chapter discusses two such applications, one pertaining to analog signal generation and the other pertaining to analog computation. As analog computing makes a resurgence, many more analog applications of straintronics are expected to appear.

8.1   Straintronic Microwave Oscillator

Low power nanoscale microwave oscillators have myriad applications: local oscillators in modulators and demodulators for FM radio, microwave assisted writer of data in magnetic memory cells (microwave assisted magnetic recording, or MAMR [1]) and coupled oscillator unit for neuromorphic computing [2]. A popular and widely researched microwave oscillator based on magneto-dynamics is the *spin-torque-* and the *spin-Hall-nano-oscillator* (STNO/SHNO) [3–5] which can now exhibit very high quality factors, up to ~10,000 [6–9], but typically outputs low power in the range of few tens to perhaps 100 nW.

Magnetic straintronics can implement a different genre of microwave (X-band) oscillators capable of producing much higher output power than an STNO (~mW), albeit with relatively low quality factor less than 100. A *single straintronic* MTJ is needed to realize the oscillator whose operation is based on the interplay between strain anisotropy, shape anisotropy, dipolar magnetic field, and spin transfer torque generated by the passage of spin polarized current through the soft layer of the MTJ [10]. Its disadvantages are the relatively low quality factor (not an issue for MAMR) and the inability to tune the frequency of the oscillation easily with the power supply voltage, which would make it unsuitable for coupled oscillator based neuromorphic computing.

Figure 8.1 shows how the device is implemented. There is some remanent dipole coupling between the hard and the soft layers of the MTJ, which is critical for device operation.

When no current passes through the MTJ (i. e. the voltage source is absent), no spin transfer torque acts on the soft layer, and no voltage is dropped over the piezoelectric layer to generate and exert strain on the soft layer. In this condition, the magnetizations of the hard and soft layers will be mutually antiparallel owing to the residual dipole coupling between them. This will place the MTJ in the *high resistance state*.

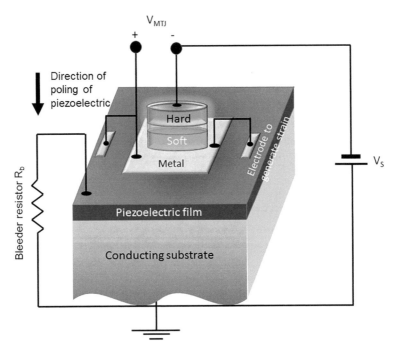

Fig. 8.1 A microwave oscillator implemented with a straintronic magneto-tunneling junction (MTJ) and a passive bleeder resistor. The bleeder resistor's resistance is much smaller than that of the piezoelectric film. The output voltage of the device is V_{MTJ} which is the voltage dropped over the MTJ. The strain generated in the elliptical MTJ soft layer due to the voltage dropped over the piezoelectric is biaxial (compressive along the major axis and tensile along the minor axis). The piezoelectric layer is poled in the vertically down direction. The soft layer is assumed to have positive magnetostriction. The lateral dimension of the soft layer, the spacing between the edge of the soft layer and the nearest electrode, and the piezoelectric film thickness are all approximately the same and that generates biaxial strain in the soft layer. Adapted from M. A. Abeed, J. L. Drobitch and S. Bandyopadhyay, *Phys. Rev. Appl.*, **11**, 054,069 (2019) with permission of the American Physical Society. © 2019 American Physical Society

If one now turns on the voltage supply V_s (with the polarity shown in Fig. 8.1), spin-polarized electrons will be injected from the hard layer into the soft layer and they will gradually turn the soft layer's magnetization in the direction of the hard layer's magnetization because of the generated spin transfer torque. This will take the MTJ toward the *low resistance state.*

Note that the voltage dropped over the piezoelectric film is

$$V_{piezo} = \frac{R_{piezo} \| R_b}{R_{piezo} \| R_b + R_{MTJ}} V_s \approx \frac{R_b}{R_b + R_{MTJ}} V_s$$

where R_{piezo} is the resistance of the piezoelectric film (the resistance of the conducting substrate is negligible and hence ignored in this discussion), R_b is the resistance of a current bleeder resistor in parallel with the piezoelectric film as shown in Fig. 8.1 and R_{MTJ} is the resistance of the MTJ. We assume that $R_b \ll R_{piezo}$.

The last equation shows that when the MTJ goes into the low resistance state ($R_{MTJ} \to$ low), the voltage dropped over the piezoelectric film V_{piezo} increases and that strains the soft layer sufficiently to rotate its magnetization away from the major axis toward the minor axis as long as the product of the strain and magnetostriction coefficient is negative (i.e. strain will have to be compressive if the magnetostriction coefficient of the soft layer is positive and vice versa). This rotation, which increases the angle between the magnetizations of the hard and soft layer, will increase the resistance of the MTJ and reduce the spin polarized current flowing through it (for a constant supply voltage V_s). At that point, the dipole coupling effect can overcome the reduced spin transfer torque associated with the reduced current passing through the MTJ and coax the soft layer's magnetization to point opposite to that of the hard layer, causing the MTJ resistance to approach the high resistance state. Once that happens, the voltage dropped over the piezoelectric film falls (see the last equation) and the strain in the soft layer falls with it. However, the spin polarized current continues to flow through the soft magnet, and over sufficient time, will transfer enough torque to the soft layer's magnetization to make it once again attempt to align parallel to the hard layer's magnetization. This will tend to take the MTJ back to the low resistance state and the process repeats itself. The MTJ resistance R_{MTJ} therefore continuously oscillates between the high and low states. This will make the voltage V_{MTJ} dropped over the MTJ $\left[V_{MTJ} = V_s R_{MTJ} / (R_{piezo} \| R_b + R_{MTJ}) \right] \approx V_s R_{MTJ} / (R_b + R_{MTJ})$ also continuously oscillate between two values, resulting in an oscillator. The reader will understand that the MTJ resistance cannot be much larger than the bleeder resistor's resistance; otherwise the voltage across the MTJ will be constant and not oscillate in time.

Figure 8.2 shows the simulated oscillation waveform for specific values of R_b, R_{AP}, R_P and V_s reported in ref. [10]. The simulation was based on the stochastic Landau-Lifshitz-Gilbert equation, which takes the effect of thermal noise into account. The inset shows the Fourier transform of the oscillations. The output power was calculated to be tens of mW for the parameters (R_b, R_{AP}, R_P and V_s) used in ref. [10].

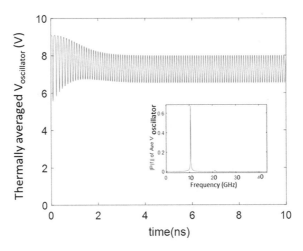

Fig. 8.2 Thermally averaged oscillator waveform at 300 K in the presence of thermal noise. This plot was obtained by averaging 1000 magneto-dynamics trajectories (evolution of the soft layer's magnetization with time under 1000 instances of random thermal noise) which are all slightly different from each other because of the randomness of the noise field. It takes about 3 ns to reach steady state amplitude. The steady state period is ~100 ps (frequency = 10.52 GHz, wavelength = 3 cm). The dc offset is about 7.3 V and the steady state peak-to-peak amplitude is 1.5 V. The inset shows the Fourier spectra of the oscillations after suppressing the dc component. The fundamental frequency is 10.52 GHz and there is a second harmonic at ~21 GHz whose amplitude is ~60 times less than that of the fundamental. Surprisingly the output is spectrally pure and this is almost a monochromatic (ideal) oscillator. The resonant frequency is 10.52 GHz and the bandwidth (full width at half maximum) is ~200 MHz, leading to a quality factor of 52.6. Reprinted from M. A. Abeed, J. L. Drobitch and S. Bandyopadhyay, *Phys. Rev. Appl.*, **11**, 054,069 (2019) with permission of the American Physical Society. © 2019 American Physical Society

8.2   Straintronic Analog Multiplier

A critical component of any analog computer is an analog multiplier. Reference [11] proposed a straintronic implementation using a single magnetic tunnel junction.

A schematic of the multiplier is shown in Fig. 8.3. Because of dipole interaction with the hard layer, the soft layer experiences an effective magnetic field H_d that is directed along its major axis opposite to the magnetization of the hard layer.

When a (gate) voltage V_G of the correct polarity is applied to the shorted electrode pair in Fig. 8.3a, it generates biaxial strain in the piezoelectric film pinched between the two electrodes, which will tend to rotate the soft layer's magnetization away from the major axis of the ellipse (the easy axis) towards the minor axis (hard axis). This is countered by the magnetic field H_d which prefers to keep the magnetization in its initial orientation. Hence, the magnetization ultimately settles into a steady-state orientation that subtends

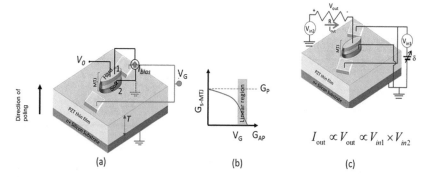

Fig. 8.3 **a** Straintronic magnetic tunnel junction configured to produce a linear region in the transfer characteristic G_{MTJ} (magnetic tunnel junction conductance) versus V_G (gate voltage). **b** The transfer characteristic showing the linear region. **c** An analog multiplier implemented with a single straintronic MTJ. The two operands are encoded in V_{in1} and V_{in2} and the product of them is encoded in V_{out} or I_{out}. The MTJ is biased in the linear region of the transfer characteristic where the MTJ conductance is proportional to $(V_G - \delta)$ with δ being a bias voltage. Reproduced from https://doi.org/10.36227/techrxiv.16649803.v1. Figures **(a)** and **(c)** are reproduced from *Nanotechnology*, **29**, 442,001 (2018)

some angle θ_{ss} with the major axis (or the magnetization of the hard layer). This is the location of the potential energy minimum. The value of θ_{ss} depends on the applied strain and H_d. Because the hard layer's magnetization is stiff, H_d is fixed and does not change with strain or the gate voltage V_G. Therefore, as one sweeps V_G (and the resulting strain), one will change θ_{ss} and hence the resistance of the MTJ which depends on θ_{ss}.

To implement the multiplier, a constant current source I_{bias} is connected between the hard and soft layers of the MTJ (terminals '1' and '2' in Fig. 8.3a) to drive a current through the MTJ. The gate voltage V_G is applied at terminal '3' to generate the strain in the soft layer and change the MTJ resistance, while a fourth terminal is connected to the hard layer (common with terminal '1'), which outputs a voltage V_0. Terminal 2, connected to the soft layer, is grounded and hence $V_0 = R_{MTJ} I_{bias}$, where R_{MTJ} is the resistance of the MTJ that can be altered by the gate voltage V_G, as explained before.

Reference [11] modeled the rotation of the soft layer's magnetization as a function of the strain (and hence the gate voltage V_G) in the presence of H_d and thermal noise using stochastic Landau-Lifshitz-Gilbert simulations. This yielded the θ_{ss} versus V_G relation. The MTJ resistance is given by $R_{MTJ} = R_P + \frac{R_{AP} - R_P}{2}[1 - \cos\theta_{ss}]$, where, R_P is the parallel resistance and R_{AP} the antiparallel one. Hence, from the θ_{ss} versus V_G relation, Ref. [11] calculated the $1/R_{s\text{-}MTJ}$ $(= G_{s\text{-}MTJ})$ versus V_G characteristic, which is shown qualitatively in Fig. 8.3b. What was remarkable was that with proper choice of the MTJ parameters, we could produce a *linear* region in the $G_{s\text{-}MTJ}$ vs. V_G characteristic where $1/R_{s\text{-}MTJ} = 1/R_{AP} + \kappa(V_G - \delta) \Rightarrow G_{s\text{-}MTJ} = G_{AP} + \kappa(V_G - \delta)$ [κ and δ are constants]. The existence of this linear region was proved *analytically* in ref. [11].

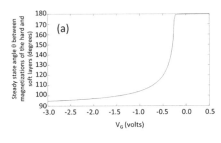
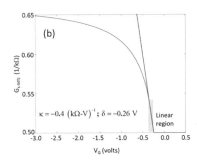

Fig. 8.4 Plots of **a** the steady-state value of the angle θ between the magnetizations of the hard and soft layers of the MTJ as a function of the gate voltage V_G obtained from the stochastic Landau-Lifshitz-Gilbert simulation at room temperature (300 K). Because of thermal noise, which introduces randomness in the magnetization trajectory, this curve was obtained by averaging over 100 trajectories. **b** The $1/R_{MTJ}$ versus V_G characteristic showing that there is a region (shaded in the figure) where the relation $G_{s-MTJ} = G_{AP} + \kappa(V_G - \delta)$ holds approximately. This plot was obtained by assuming $R_P = 1$ kΩ and $R_{AP} = 2$ kΩ. The voltage δ and the constant κ obtained by fitting a straight line to this plot are mentioned in the figure. These values were: $\kappa = -0.4 \pm 0.045$ (kΩ-V)$^{-1}$ and δ $= -0.26 \pm 0.013$ V. Reproduced from https://doi.org/10.36227/techrxiv.16649803.v1

Figure 8.4a shows the θ_{ss} versus V_G characteristics obtained from the stochastic Landau Lifshitz Gilbert simulation and the resulting G_{MTJ} versus V_G plot obtained for certain choices of the MTJ parameters. For these choices, listed in Ref. [11], there is indeed a region of V_G where $1/R_{s-MTJ} = 1/R_{AP} + \kappa(V_G - \delta) \Rightarrow G_{s-MTJ} = G_{AP} + \kappa(V_G - \delta)$. This is shown in Fig. 8.4b. When the gate voltage V_G is chosen to be in that region, one can perform an analog multiplication of two input voltages V_{in1} and V_{in2} encoding the multiplier and multiplicand, as explained next.

In Fig. 8.3c, the configuration is set up so that $V_{in1} = V_G - \delta$. If V_G is within the linear region in Fig. 8.4b, then $G_{MTJ} = G_{AP} + \kappa(V_G - \delta) = G_{AP} + \kappa V_{in1}$. The voltage dropped over the series resistor R in Fig. 8.3c is

$$V_{\text{out}} = I_{\text{out}} R = \frac{R}{R + R_{MTJ}} V_{in2} \approx \frac{R}{R_{MTJ}} V_{in2} \approx R G_{MTJ} V_{in2} \text{ if } R \ll R_{MTJ}$$

Replacing G_{MTJ} in the preceding Equation with $G_{AP} + \kappa V_{in1}$, one obtains

$$V_{\text{out}} = R G_{AP} V_{in2} + R\kappa(V_{in1} \times V_{in2}) \approx R\kappa(V_{in1} \times V_{in2}) \text{ and}$$
$$I_{\text{out}} \approx \kappa(V_{in1} \times V_{in2}) \text{ since } R \ll R_{AP}$$

That implements a "multiplier" since the current I_{out} flowing through the MTJ (which is also the current through the series resistor R) is proportional to the *product* of the two input voltages V_{in1} and V_{in2}. The voltage V_{out} is proportional to this current and hence it too is proportional to the product $V_{in1} \times V_{in2}$. Similar ideas were used to design probability composer circuits for Bayesian inference engines in the past [12].

References

1. S. Okamoto, N. Kikuchi, M. Furuta, O. Kitakami, T. Shimatsu, Microwave assisted magnetic recording technologies and related physics. J. Phys. D: Appl. Phys. **48**, 353001 (2015)
2. J. Torrejon, M. Riou, A.A. Flavio, S. Tsunegi, G. Khalsa, D. Querlioz, P. Bortolotti, V. Cros, A. Kukushima, H. Kubota, S. Yuasa, M.D. Stiles, J. Grollier, Neuromorphic computing with nanoscale spintronic oscillators. Nature **547**, 428 (2017)
3. A.M. Deac, A. Fukushima, H. Kubota, H. Maehara, Y. Suzuki, S. Yuasa, Y. Nagamine, K. Tsunekawa, D.D. Djayaprawira, N. Watanabe, Bias-driven large power microwave emission from MgO-based tunnel magnetoresistance devices. Nature Phys. **4**, 803 (2008)
4. T. Chen, R.K. Dumas, A. Eklund, P.K. Muduli, A. Houshang, A.A. Awad, P. Dürrenfeld, M.G. Malm, A. Rusu, J. Åkerman, Spin torque and spin Hall nano-oscillators. Proc. IEEE **104**, 1919 (2016)
5. Z. Zeng, P.K. Amiri, I.N. Krivorotov, H. Zhao, G. Finocchio, J.-P. Wang, J. Katine, Y. Huai, J. Langer, K. Galatsis, K.L. Wang, H.W. Jiang, High power coherent microwave emission from magnetic tunnel junction nano-oscillators with perpendicular anisotropy. ACS Nano **6**, 6115 (2012)
6. S. Jiang, R. Khymyn, S. Chung, T.Q. Le, L.H. Diez, A. Houshang, M. Zahedinejad, D. Ravelosona, J. Åkerman, Reduced spin torque nano oscillator linewidth using He$^+$ irradiation. Appl. Phys. Lett. **116**, 072403 (2020)
7. H. Maehara, H. Kubota, Y. Suzuki, T. Seki, K. Nishimura, Y. Nagamine, K. Tsunekawa, A. Fukushima, H. Arai, T. Taniguchi, H. Imamura, K. Ando, S. Yuasa, High Q factor over 3000 due to out of plane precession in nano-contact spin torque oscillator based on magnetic tunnel junctions. Appl. Phys. Express **7**, 023003 (2014)
8. D. Kumar, K. Konishi, N. Kumar. S. Miwa, A. Fukushima, K. Yakushiji, S. Yuasa, H. Kubota, C.V. Tomy, A. Prabhakar, Y. Suzuki, A. Tulapurkar, Coherent microwave generation by spintronic feedback oscillator. Sci. Rep. **6**, 30747 (2016)
9. H. Fulara, M. Zahedinejad, R. Khymyn, A.A. Awad, S. Muralidhar, M. Dvornik, J. Akerman Spin wave torque driven propagating spin waves. Sci. Adv. **5**, eaax8467 (2019)
10. M.A. Abeed, J.L. Drobitch, S. Bandyopadhyay, Microwave oscillator based on a single straintronic magnetotunneling junction. Phys. Rev. Appl. **11**, 054069 (2019)
11. S. Bandyopadhyay, R. Rahman, A non-volatile all-spin analog matrix multiplier: An efficient hardware accelerator for machine learning. https://doi.org/10.36227/techrxiv.16649803.v1
12. S. Khasanvis, M. Li, M. Rahman, M. Salehi-Fashami, A.K. Biswas, J. Atulasimha, S. Bandyopadhyay, C.A. Moritz, Self-similar magneto-electric nanocircuit technology for probabilistic inference engines. IEEE Trans. Nanotechnol. **14**, 980 (2015)

Straintronic Nano-Antennas

9

Antennas are used in all manner of communications—from cell phones, to self-driving automobiles, to television signal reception, to drones, to medically implanted devices that must communicate with external monitors. One debilitating drawback of conventional electromagnetic antennas is that when they are aggressively miniaturized to dimensions much smaller than the wavelength of the radiation they emit, their *gain* and *radiation efficiency* plummet. The theoretical limits on these quantities for conventional antennas are, respectively, A/λ^2 and $A/(2\pi\lambda)^2 + \left(\sqrt{A}\Big/\pi\lambda\right)$ [1], where A is the emitting area of the antenna and λ is the wavelength of the radiated wave. If the emitting area is two orders of magnitude smaller than the square of the wavelength, then the radiation efficiency is limited to 1% and the gain to 0.03. This hinders embedded applications where the antenna has to be made orders of magnitude smaller than the wavelength (at radio frequencies) in order to be integrated on-chip, integrated into wearable electronics, or medically implanted inside a patient's body.

A new genre of antennas have recently attracted attention because they are very different from conventional antennas and their radiation efficiency and gain are *not constrained* by the above theoretical limits. Primary among them are magneto-elastic antennas or *straintronic antennas* [2–5]. Conventional antennas are activated by passing a time varying current (or electromagnetic wave) through a metallic or non-metallic element which gives rise to time-varying charges on these elements. These charges radiate an electromagnetic wave with the same frequency as the excitation. A straintronic antenna, on the other hand, works very differently. It is implemented with an array of magnetostrictive nanomagnets fabricated on a piezoelectric substrate and periodically strained by a (surface) acoustic wave launched in the substrate. The periodic strain makes the magnetizations within the nanomagnets precess about their magnetic easy axis. These time-varying magnetizations, rather than time-varying charges, radiate electromagnetic waves with the same frequency as the acoustic wave and thus implement an antenna.

© The Author(s), under exclusive license to Springer Nature Switzerland AG 2022   95
S. Bandyopadhyay, *Magnetic Straintronics*, Synthesis Lectures on Engineering, Science,
and Technology, https://doi.org/10.1007/978-3-031-20683-2_9

The major differences that a straintronic antenna has with a conventional antenna are that: (1) here, instead of an electromagnetic wave, an acoustic wave acts as the input signal, and (2) the radiating elements are not oscillating changes but oscillating magnetizations. Because of these differences, a straintronic antenna does not suffer from the same limitations on radiation efficiency and gain that conventional antennas suffer from. It has been experimentally shown that extreme sub-wavelength straintronic antennas $\left(A/\lambda^2 \ll 1 \right)$ can have gains and efficiencies that exceed the theoretical limits for conventional antennas of the same size by several orders of magnitude. This is a remarkable result that raises hopes of being able to miniaturize antennas aggressively for embedded applications.

In the *low frequency limit* (<1 GHz), straintronic antennas work essentially via the Villari effect. A low frequency surface acoustic wave gives rise to low frequency time-varying strain that induces low frequency magnetization precession in a magnetostrictive nanomagnet which radiates electromagnetic waves owing to the magnetization varying periodically in time [6]. This mechanism, however, cannot work at high frequencies since the precession cannot keep up with the rapid strain variation if the frequency is too high. At *high frequencies* (>1 GHz), a different mechanism may take over. The high frequency time-varying strain can give rise to *resonant* spin waves inside a magnetostrictive nanomagnet owing to magneto-elastic (phonon-magnon) coupling. These are standing spin waves inside the nanomagnet, which radiate electromagnetic waves because of magnon-photon coupling. This mechanism cannot be operative at low frequencies since at low frequencies, the wavelength of the spin wave may be much larger than the dimension of the nanomagnet and hence half a wavelength cannot fit inside a nanomagnet, which would be the minimum requirement for resonance. Thus, two slightly different mechanisms undergird the operation of spintronic antennas at low (radio frequency) and high (microwave) frequencies. Regardless of which mechanism is at play, straintronic antennas can exhibit gains and radiation efficiencies unattainable in conventional antennas of the same size. This is what makes them attractive.

9.1   Straintronic RF Electromagnetic Antennas Actuated by the Inverse Magnetostrictive (Villari) Effect

Consider the structure shown in Fig. 9.1 consisting of a two-dimensional periodic array of magnetostrictive nanomagnets deposited on a piezoelectric substrate between two contact pads delineated on the piezoelectric surface. A time-varying periodic voltage applied between the contact pads generates a time-varying periodic strain in the piezoelectric, which is transferred partially or wholly to the nanomagnets. This periodically deflects the magnetization of a magnetostrictive nanomagnet from the easy axis (owing to the Villari effect), causing the magnetization to vary with time with a frequency *equal to that of the signal as long as the frequency is low enough to allow the magnetization to keep*

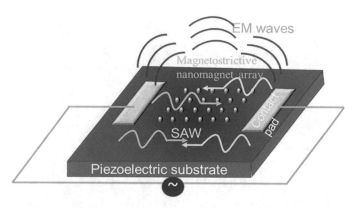

Fig. 9.1 Schematic of a low-frequency straintronic antenna based on the inverse magnetostriction (Villari) effect. A time-varying voltage applied between the contact pads generates a time-varying strain in the piezoelectric and hence in the nanomagnets. This strain can periodically deflect the magnetization of a nanomagnet from the easy axis with the frequency of the applied voltage as long as the frequency is low enough to allow the magnetization to respond quasi-statically to the strain. The time-varying magnetization associated with the periodic deflection will radiate an electromagnetic wave of the signal frequency, thus implementing a straintronic antenna. Reproduced from *Adv. Mater. Technol.*, **5**, 2,000,316 (2020) with permission of Wiley

up with the signal. If the frequency is too high, the deflections cannot keep up with the signal and hence the magnetization will not deflect (or deflect very little). At low signal frequencies however, there can be sufficient deflection, and hence the time varying magnetizations associated with the deflection can radiate electromagnetic waves at the signal frequency, thereby realizing a low frequency "straintronic" antenna. A radio frequency antenna based on this principle was demonstrated in Ref. [5]. It could radiate at 144 MHz, but not 900 MHz, because the latter frequency was too high. The antenna was several orders of magnitude smaller than the wavelength at 144 MHz ($A/\lambda^2 = 10^{-8}$) and yet it radiated with efficiency $> 10^{-3}$ which was 144,000 times larger than A/λ^2.

It might appear from the preceding discussion that these straintronic antennas can only work at low frequencies, but that is not true. Suppose that the frequency is high, but it is the ferromagnetic resonance (FMR) frequency of the nanomagnets. In this case, even the small deflections caused by the high frequency strain will build up coherently over a few cycles, as discussed previously in Sect. 7.3, resulting ultimately in a large deflection that can radiate an electromagnetic wave at the FMR frequency with sufficient power. A relatively high frequency antenna based on this principle was demonstrated in Ref. [3]. Its radiation efficiency also exceeded the theoretical A/λ^2 limit for conventional antennas.

Figure 9.2 shows scanning electron micrographs of the Co nanomagnets used for the antenna in Ref. [5]. Note that the array does not have rotational symmetry since the major axis dimension is different from the minor axis dimension of any individual nanomagnet, and the edge-to-edge spacing between the nanomagnets is also different in the two directions (along the major axis and along the minor axis). Hence, the radiation pattern of

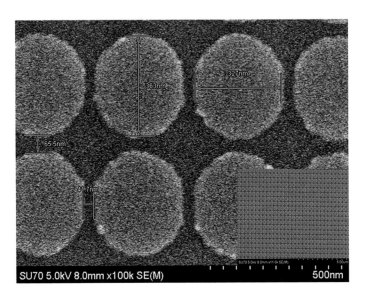

Fig. 9.2 Scanning electron micrographs of the nanomagnet array used to demonstrate an ultra-sub-wavelength RF straintronic antenna ($A/\lambda^2 = 10^{-8}$) working at 144 MHz with a radiation efficiency 144,000 times larger than A/λ^2. The inset shows a low-magnification image of the array. Reproduced from *Adv. Mater. Technol.*, **5**, 2,000,316 (2020) with permission of Wiley

these antennas may not be completely isotropic even though the antennas may appear to be point sources since their dimensions are so much smaller than the wavelength. One may be also able to alter the radiation pattern by placing the electrodes generating the strain at a different location so that the strain wave propagation direction is altered. This raises the tantalizing possibility of beam steering by activating different sets of electrodes and hence an active electronically scanned antenna (AESA). These are areas of current research.

9.2   Straintronic Microwave Electromagnetic Antennas Actuated by Tripartite Phonon-Magnon-Photon Coupling

A high-frequency (microwave) sub-wavelength straintronic antenna was implemented with a modality somewhat different from the one discussed in Sect. 9.2. Consider the structure in Fig. 9.3b, where a time-varying voltage (of several GHz frequency) applied between electrode pairs (3,4) or (5,7) will launch a surface acoustic wave (SAW) of ~GHz frequency in the piezoelectric LiNbO3 substrate. Depending on which electrode pair is activated, the SAW will propagate primarily in two mutually orthogonal directions. These SAWs excite confined spin waves (of the same frequency as the SAW) in the nanomagnets

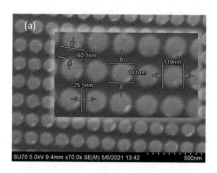 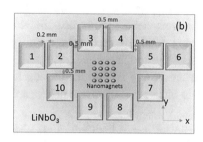

Fig. 9.3 **a** Scanning electron micrograph of elliptical Co nanomagnets deposited on a $LiNbO_3$ substrate. The inset shows the dimensions of the nanomagnets and edge-to-edge separation between nearest neighbors. **b** The antenna configuration with electrodes for launching surface acoustic waves that subject the nanomagnets to time-varying strain at the frequency of the launched wave. The dimensions are shown. This figure is not to scale. Reproduced from *Adv. Sci.*, **9**, 2,104,644 (2022) with permission of Wiley

via the coupling of the phonons in the SAW to magnons in the spin wave [7]. The generated spin waves can then couple to electromagnetic radiative modes via magnon-photon coupling, thereby radiating electromagnetic waves into the surrounding medium. This implements an electromagnetic antenna via tripartite phonon-magnon-photon coupling [4].

High radiation efficiencies and antenna gain (exceeding the theoretical limits on these quantities for conventional antennas of the same size by several orders of magnitude) were observed in such an antenna at microwave frequencies where three conditions were satisfied: (1) energy could be coupled efficiently from the microwave voltage source into the SAW in the substrate owing to good impedance matching, (2) the launched SAW was able to excite *resonant* spin waves in the nanomagnets because of relatively efficient phonon-magnon coupling, and (3) the resonant spin waves could couple somewhat efficiently into radiative electromagnetic modes because of a reasonable degree of magnon-photon coupling. Thus, three-way coupling between phonons, magnons and photons is required. Figure 9.4 shows the measured radiation efficiency and gain for two different directions of SAW propagation—one parallel to the major axes and the other parallel to the minor axes of the elliptical nanomagnets—in the antenna shown in Fig. 9.3 [6]. These measured quantities were compared with the theoretical limits for conventional antennas of the same emitting areas to showcase the fact that they exceed those limits by several orders of magnitude. This is shown in Fig. 9.4. Interestingly, both the gain and the efficiency, as well as the spin wave profiles in space, depend on the direction of propagation of the SAW with respect to the nanomagnet geometry, i.e. whether the propagation is along the major or minor axes of the nanomagnets. This is not surprising since the major axis is the magnetic easy axis and the minor axis is the magnetic hard axis. Additionally, the edge-to-edge separation between the nanomagnets (and hence the degree of dipole coupling between neighbors) is also different along the two directions, which exacerbates the anisotropy. Similar anisotropy effects were observed and studied in Ref. [7] in a system that was very similar to the straintronic antenna shown in Fig. 9.3.

The anisotropy is also manifested in the spectrum of the scattering parameter S_{11}, which is different for the two different directions of SAW propagation. This is shown in Fig. 9.5.

The anisotropy is both intriguing and enticing since it raises the specter of being able to alter the radiation pattern by activating different electrode pairs in Fig. 9.3b. If different pairs are activated sequentially with a specific phase difference, one may be able to steer

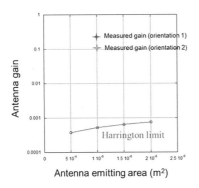
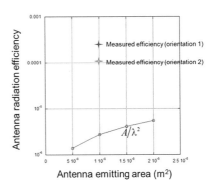

Fig. 9.4 Plots of antenna gain and radiation efficiency as a function of the antenna emitting area at frequency of 5 GHz (wavelength 6 cm), showing the theoretical limits and the measured quantities in both orientations. Orientation 1 corresponds to the SAW propagating along the major axes and orientation 2 corresponds to the SAW propagating along the minor axes of the elliptical nanomagnets. Reproduced from *Adv. Sci.*, **9**, 2,104,644 (2022) with permission of Wiley

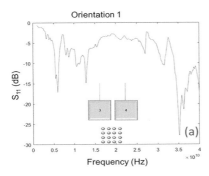
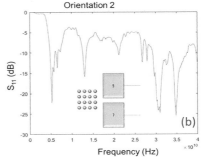

Fig. 9.5 The spectra of the scattering parameter S11. **a** The surface acoustic wave travels parallel to the minor axes of the elliptical nanomagnets, and **b** the surface acoustic wave travels parallel to the major axes of the nanomagnets. Reproduced from *Adv. Sci.*, **9**, 2,104,644 (2022) with permission of Wiley

the radiated beam and that can implement an actively electronically scanned antenna (AESA). In this case, the antenna dimension is much smaller than the radiated wavelength and hence any such implementation will enable embedded AESAs, which could be a disruptive technology.

To further emphasize the anisotropy feature, the simulated spin wave profiles associated with oscillations of the magnetization component parallel to the minor axis of the elliptical nanomagnets, as well as the radiation spectra, are shown in Figs. 9.6 and 9.7 for two cases when the frequency of the launched SAW was 5 GHz. The two cases correspond to the SAW propagating along the major axis and along the minor axis of the nanomagnets. In the former case, the dominant peak in the radiation spectra is at the 5 GHz excitation frequency as expected, but in the latter case, it is surprisingly not at the excitation frequency but at a sub-harmonic of 2.5 GHz. This happens because here the 5 GHz SAW excites a spin wave most efficiently not at its own frequency, but at one-half its own frequency.

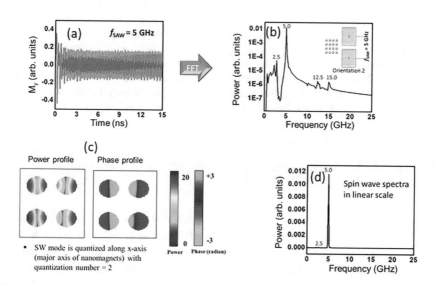

Fig. 9.6 **a** Calculated oscillations in the magnetization component parallel to the minor axis of the elliptical nanomagnets (hard axis) when the exciting SAW frequency is 5 GHz. The sample is excited by activating two electrodes that make the SAW propagate along the ***major axes*** of the elliptical nanomagnets. **b** The power spectrum of the oscillations in log linear scale where the dominant peak is at the excitation frequency. The next dominant peak has a power amplitude more than 2 orders of magnitude smaller and is at 2.5 GHz. **c** The power and phase profiles of the spin waves in the 5 GHz (dominant frequency) mode showing that the wave is quantized along the major axis (which is also the direction of propagation of the SAW) with quantization number = 2. **d** The power spectrum of the oscillations in linear scale. Reproduced from *Adv. Sci.*, **9**, 2,104,644 (2022) with permission from Wiley

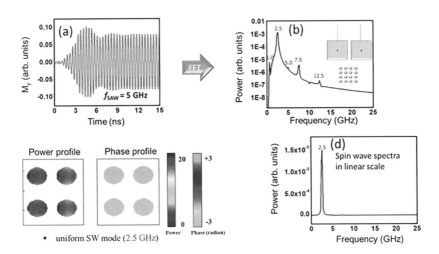

Fig. 9.7 a Calculated oscillations in the magnetization component parallel to the minor axis of the elliptical nanomagnets when the exciting SAW frequency is 5 GHz. The sample is excited by activating two electrodes that make the SAW propagate along the ***minor axes*** of the elliptical nanomagnets. **b** The power spectrum of the oscillations in log linear scale where the dominant peak is *not* at the excitation frequency of 5 GHz, but at the sub-harmonic of 2.5 GHz. There is also a lower amplitude peak at 5 GHz whose power is two orders of magnitude smaller. **c** The power and phase profiles of the spin waves in the 2.5 GHz (dominant frequency) mode showing that the wave is uniform. **d** The power spectrum of the oscillations in linear scale. Reproduced from *Adv. Sci.*, **9**, 2,104,644 (2022) with permission from Wiley

9.3   Acoustic Nano-Antennas Actuated by the Direct Magnetostrictive Effect

Section 9.1 dealt with *electromagnetic* antennas actuated by the *inverse magnetostriction* effect. This section deals with *acoustic* antennas that can be actuated by the *direct magnetostriction* effect. An acoustic antenna radiates an acoustic wave into a surrounding solid and hence can be used to launch a surface acoustic wave (SAW) in a substrate much like an interdigitated transducer (IDT). The straintronic type that we will discuss here however has one significant difference with an IDT; whereas an IDT will have dimensions typically much larger than the acoustic wavelength, this one can have dimensions that are orders of magnitude smaller than the acoustic wavelength and yet radiate an acoustic wave efficiently. A second difference is that an IDT can launch a SAW at only one frequency (its own resonant frequency) or a narrow range of frequencies. The straintronic acoustic antenna discussed here can launch a surface acoustic wave at any arbitrary frequency as long as the frequency is low enough to allow time varying strain to reverse the magnetizations of the magnetostrictive nanomagnets that make up this antenna. Yet a third

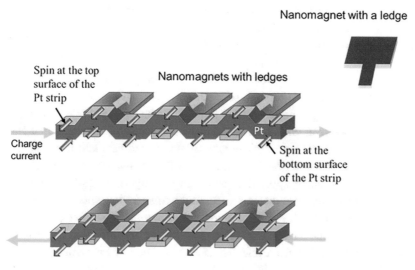

Fig. 9.8 Principle of actuation of the ultra-sub-wavelength acoustic antenna by alternating spin current injection from the heavy metal (Pt) nanostrip. For one polarity of the injected charge current into the nanostrip, the domain wall in the nanomagnet moves in one direction and for the opposite polarity, it moves in the opposite direction. This makes the magnetizations oscillate periodically with the frequency of the spin (or charge) current. The nanomagnets expand and contract as their magnetizations oscillate (because they are magnetostrictive) and that generates a surface acoustic wave in the substrate which can be detected by interdigitated transducers delineated on the substrate if the substrate is piezoelectric. Reproduced from *Adv. Mater. Technol.*, **5**, 1,901,076 (2020) with permission of Wiley

difference is that an IDT can launch a SAW in a piezoelectric substrate, whereas this antenna can launch a SAW in any substrate, whether or not it is piezoelectric.

A straintronic acoustic antenna was realized by harnessing the *giant spin Hall effect* in heavy metals like platinum. The input signal was an alternating charge current that was passed through a heavy metal (Pt) strip which overlies "ledges" in magnetostrictive nanomagnets deposited on a substrate (see Fig. 9.8). Thus, the strip and the nanomagnets are in "magnetic contact" which allows spins to diffuse from the strip into the nanomagnets through the ledges and induce domain wall motion in the nanomagnets. During the positive cycle of the alternating charge current, the domain wall will move in one direction and magnetize the nanomagnets in that direction, while during the negative cycle, the domain wall will move in the opposite direction and magnetize the nanomagnets in the opposite direction. The frequency of the current has to be of course low enough to allow the domain wall to move completely through the magnet during either cycle and thus allow the magnetization to alternate with the period of the charge current. Since the nanomagnets are magnetostrictive, they will expand and contract during each cycle, thereby generating a periodic strain in the substrate underneath, which constitutes a surface acoustic wave. The system is designed such that the bulk of the nanomagnets lies outside the strip and only the ledges are in contact with the strip to allow spin diffusion from the strip into the nanomagnets. This ensures that the strip does not "clamp" the nanomagnets and hence does not prevent them from expanding and contracting periodically in response to the alternating current. A schematic is shown in Fig. 9.8.

The alternating charge current passing through the heavy metal injects spin current of *alternating* spin polarization into the nanomagnets because of the giant spin Hall effect in Pt. The injected spins cause domain wall motion in the nanomagnets. The distance a domain wall moves in a duration Δt (within either the positive or the negative cycle of the current) is $x = v_d \Delta t$ where v_d is the domain wall velocity. If one half of the period of the alternating charge current significantly exceeds the time L / v_d, where L is the dimension of the nanomagnet in the direction of domain wall motion, then after completion of the positive cycle, the nanomagnet will be magnetized in one direction and subsequently, after completion of the negative cycle, the nanomagnet will be magnetized in the opposite direction. The magnetizations will therefore alternate, and because the nanomagnets are magnetostrictive, they will periodically expand and contract owing to the magnetostriction effect. This periodic expansion/contraction will generate a periodic strain in the underlying piezoelectric, leading to the propagation of a surface acoustic wave (SAW) in the substrate that can be detected as an oscillating electrical signal with interdigitated transducers (IDT) delineated on the substrate if the substrate is piezoelectric. All that is required for this to happen is that the signal frequency $f < v_d / L$. The wavelength of the generated acoustic wave is determined solely by the frequency of the alternating charge current and the velocity of acoustic wave propagation in the substrate $\left(\lambda_{ac} = v_{ac} / f \right)$. Hence, we arrive at the condition $L / \lambda_{ac} \le v_d / v_{ac}$ which will allow the nanomagnet linear dimension L to be much smaller than the acoustic wavelength λ_{ac} since the domain wall velocity is much smaller than acoustic wave velocity. Thus, this construct can be an *ultra-sub-wavelength* acoustic antenna, which an IDT cannot be.

Such an ultra-sub-wavelength acoustic antenna, 67 times smaller than the acoustic wavelength that it radiated, was demonstrated experimentally in Ref. [8]. It is shown schematically in Fig. 9.8. Figure 9.9 shows the device layout and scanning electron micrographs of various components used in the experiment to demonstrate the ultra-sub-wavelength acoustic antenna in Ref. [8]. Figure 9.10 shows oscilloscope traces of the waveforms of the alternating current injected into the Pt nanostrip, as well as the oscilloscope traces of the signals picked up at the IDTs because of the SAW generated in the piezoelectric substrate. The experiment used a frequency of ~3.5 MHz for the injected current and the generated SAW was of the same frequency.

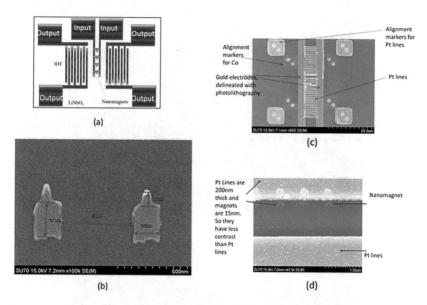

Fig. 9.9 **a** Pattern for the acoustic antenna showing the interdigitated transducers for detecting a SAW generated in the piezoelectric LiNbO₃ substrate, the heavy metal (Pt) strips (in chrome yellow) and the magnetostrictive nanomagnets (made of Co) in contact with the heavy metal strips. This figure is not to scale. **b** Scanning electron micrographs of the fabricated Co nanomagnets with Gaussian shaped ledges. **c** Scanning electron micrograph of the Pt lines overlying the nanomagnets. **d** Zoomed view showing the nanomagnets underneath the Pt line. Reproduced from *Adv. Mater. Technol.*, **5**, 1,901,076 (2020) with permission of Wiley

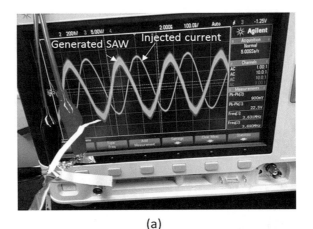

(a)

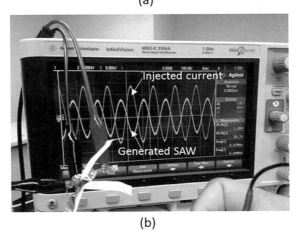

(b)

Fig. 9.10 Oscilloscope traces of the alternating voltage applied across the Pt lines to actuate the acoustic antenna and the alternating voltage detected at the interdigitated transducer due to the SAW radiated by the acoustic antenna. **a** The frequency of the voltage across the Pt lines is 3.63 MHz and the peak-to-peak amplitude is 22.5 V. The Pt lines are about 200 nm thick, 1 μm wide and 10 μm long, which makes their resistance about 5 ohms. Hence the peak-to-peak current that is passed through them is 4.5 A and the resulting current density is $J_c = 2.25 \times 10^{13}$ A-m^{-2}. Assuming a spin Hall angle of 0.1, the peak to peak spin current density is $J_s = 2.25 \times 10^{12}$ A-m^{-2} and the rms spin current density is $J_{rms} = 1.6 \times 10^{12}$ A-m^{-2}. If we assume that the domain wall velocity at this current is 500 m/s, then the domain wall displacement in either the positive or the negative cyr-cle of the alternating current ~70 μm, which is much larger than the nanomagnet dimension. Hence, during either cycle, the nanomagnet is probably fully magnetized in one direction. The detected volt-age peak-to-peak amplitude (due to the generated SAW) is 0.9 V. **b** In this case, the input voltage frequency is 6.87 MHz and the peak-to-peak amplitude is 25.7 V, while the detected voltage peak-to-peak amplitude is 1.65 V. The phase shift between the two waveforms is due to the finite velocity of the SAW in the substrate that introduces a time lag between the current injection and detection of the SAW. Reproduced from *Adv. Mater. Technol.*, **5**, 1,901,076 (2020) with permission of Wiley

References

1. R.F. Harrington, Effect of antenna size on gain, bandwidth and efficiency. J. Res. Nat. Bur. Stand.-D Radio Propag. **64D**, 1 (1960)
2. J.P. Domann, G.P. Carman, Strain powered antennas. J. Appl. Phys. **121**, 044905 (2017)
3. T. Nan et al., Acoustically actuated ultra-compact NEMS magnetoelectric antennas. Nat. Commun. **8**, 296 (2017)
4. R. Fabiha, J. Lundquist, S. Majumder, E. Topsakal, A. Barman, S. Bandyopadhyay, Spin wave electromagnetic nano-antenna enabled by tripartite phonon-magnon-photon coupling. Adv. Sci. **9**, 2104644 (2022)
5. J.L. Drobitch, A. De, K. Dutta, P.K. Pal. A. Adhikari, A. Barman, S. Bandyopadhyay, Extreme sub-wavelength magneto-elastic electromagnetic antenna implemented with multiferroic nanomagnets. Adv. Mater. Technol. **5**, 2000316 (2020)
6. J. Donnevert, *Maxwell's Equations: From Current Density Distribution to Radiation Field of the Hertzian Dipole* (Springer Fachmedien Wiesbaden GmbH, Wiesbaden, Germany, 2020). Chapter 4
7. A. De, J.L. Drobitch, S. Majumder, S. Barman, S. Bandyopadhyay, A. Barman, Resonant amplification of intrinsic magnon modes and generation of new extrinsic modes in a two-dimensional array of interacting multiferroic nanomagnets by surface acoustic waves. Nanoscale **13**, 10016 (2021)
8. M.A. Abeed, S. Bandyopadhyay, Experimental demonstration of an extreme subwavelength nanomagnetic acoustic antenna actuated by spin-orbit torque from a heavy metal nanostrip. Adv. Mater. Technol. **5**, 1901076 (2020)

Non-Boolean Straintronic Processors

10

We had emphasized throughout this monograph that straintronic switching is extremely energy-efficient, but also error-prone. The error-resilience is probably too poor for Boolean logic and therefore it behooves us to explore non-Boolean paradigms for computing and information processing where the vulnerability to errors is not a show-stopper, but the energy efficiency and non-volatility are a boon. It is in these applications that straintronics can find a niche.

The ideal application space will be collective computation models which are much more forgiving of switching failures and errors than Boolean logic. They are much more forgiving because in these paradigms, many devices work in unison to produce the computational result, and the inability of a few devices to function correctly is not catastrophic. Collective computational models have been used frequently to solve combinatorial optimization problems. A well-known example is an Ising machine which casts the optimum solution of a combinatorial optimization problem as the ground state of a system of interacting binary devices represented by an Ising (spin) Hamiltonian [1]. Boltzmann machines are similar. Hardware implementation of Ising computers have involved CMOS [2], trapped ions [3] and electromechanical systems [4], to name a few. Other collective computational approaches to solving combinatorial optimization problems have made use of simulated annealing [5], quantum annealing [6], cellular neural networks [7], lasers [8], quantum dots/wires [9, 10] and nanomagnets [11–14].

The quintessential example of an error-resilient collective computational task is image processing which is known to be extremely forgiving of individual device errors. Reference [9] examined the working of an image processor implemented with interacting nanowires (interacting through charge exchange) and found that the circuit can perform the rudimentary image processing task even if ~30% of the constituents fail! This happens because no single nanowire is critical and the image processing activity emerges from the cooperative actions of many nanowires interacting with each other. Similar resilience against individual device malfunctions are expected if the image processing activity is elicited from an array of dipole coupled nanomagnets clocked or activated with strain.

10.1   Straintronic Image Processor

A two-dimensional periodic array of *skewed* straintronic magnetic tunnel junctions (MTJ), of elliptical shape, fabricated on a piezoelectric substrate, can perform a variety of black-and-white image processing tasks by exploiting nearest neighbor dipole coupling between their magnetostrictive soft layers. The soft layers possess in-plane magnetic anisotropy, and are in elastic contact with the piezoelectric substrate. A global in-plane magnetic field is applied along the minor axes of the soft layers to bring their stable magnetization orientations out of the major axis and lie in the nanomagnet's plane subtending an angle of 90° between themselves. This is shown in Fig. 10.1. The hard layer is permanently magnetized in the direction of one of the two stable orientations of the soft layer's magnetization. The two resistance values of the MTJ, when the soft layer's magnetization is in either stable direction, encode black and white pixel colors. Each MTJ encodes the color of a pixel and there are as many MTJs as pixels in an image.

The pixel colors are first converted to voltage states (white = positive polarity voltage; black = negative polarity voltage) with photodetectors and level shifters. The voltage for a pixel is then applied across the skewed MTJ corresponding to that pixel. When the voltage polarity is negative (black pixel), electrons are injected from the hard into the soft layer of the MTJ, which delivers a spin transfer torque (STT) and drives the soft layer into state 1 (magnetization parallel to that of the hard layer) as shown in Fig. 10.1 and writes the pixel color "black" into the MTJ resistance state (low resistance = black). When the voltage polarity is positive (corresponding to white pixel) electrons with spins aligned along the magnetization of the hard layer are extracted from the soft layer via spin-dependent tunneling through the spacer layer (shown in gray) and the latter's magnetization switches to the other stable state 2. This is the high resistance state of the MTJ which encodes the pixel color white. Thus, a pixel color is "written" into the resistance state of an MTJ and it is "read" by measuring the resistance.

To write the pixel color using strain instead of spin transfer torque, we apply a voltage to two antipodal contact pads aligned along the direction of stable state 1 which encodes the black pixel. If a positive voltage is applied, it will generate tensile strain along the line joining the contact pads (because of the direction of poling shown in Fig. 10.1) and compressive strain in the perpendicular direction. This will magnetize the soft layer in the direction of the stable state 2, writing a "white" pixel, if the magnetostriction of the soft layer is negative. Similarly, a voltage of negative polarity will write a black pixel. The write error probability is of course much higher if we write with strain instead of using STT, but the write energy dissipation will also be much less if we employ strain.

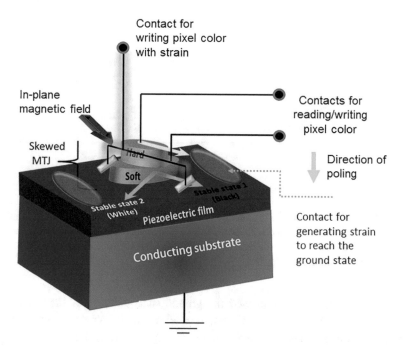

Fig. 10.1 A skewed straintronic MTJ whose soft layer has two stable magnetization orientations corresponding to black and white pixels. The two stable orientations subtend an angle of ~90° between themselves. A black pixel is written by applying a negative voltage between the hard and soft layers while a white pixel is written by applying a voltage of the opposite polarity, using spin transfer torque. To write a pixel color with strain, we apply a voltage of the appropriate polarity to the pair of contacts aligned along the direction of stable state 1. The pixel color is read by measuring the MTJ resistance. A low resistance corresponds to a black pixel and a high resistance to a white pixel. Adapted from *IEEE Trans. Elec. Dev.* **64** 2417 (2017) with permission of the Institute of Electrical and Electronics Engineers

The potential energy profile within any soft layer as a function of the angle θ that its magnetization subtends with its major axis is shown in Fig. 10.2. The profile is *asymmetric* owing to dipole interaction of the nanomagnet with its neighbors, which creates a ground state and a metastable state. The top panel in the right column shows the profile in the absence of any stress. The soft layer's magnetization may be stuck in a metastable state and not be able to reach the ground state because of the intervening potential barrier. This is depicted by the gray ball, representing the magnetization state, being trapped in the local minimum. If the barrier surrounding the local minimum is eroded by applying uniaxial stress along the major axis of the ellipse (tensile stress for a nanomagnet with negative magnetostriction and compressive stress for a nanomagnet with positive magnetostriction), then the magnetization can reach the ground state as shown in the bottom panel in the right column.

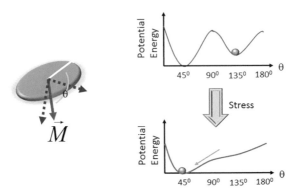

Fig. 10.2 Potential energy profile (potential energy as a function of magnetization orientation) in the soft layer of any MTJ, with and without uniaxial stress applied along the major axis of the elliptical soft layer. The sign of the stress (tensile or compressive) is chosen to make the product of the stress and the magnetostriction negative

In Chapter 7, Fig. 7.6, we showed the ground state ordering of the magnetizations in a two-dimensional array of dipole coupled elliptical nanomagnets where the edge-to-edge separation between nearest neighbors along the row and along the column are about the same. The major axes are aligned along columns and the minor axes along the rows. In this case, the ground state magnetic ordering will be ferromagnetic along columns and anti-ferromagnetic along rows. However, we can consider a different situation where the edge-to-edge separation along rows is much larger than that along columns. In this case, there is very weak dipole interaction along rows and much stronger interaction along columns. The ground state ordering will definitely be ferromagnetic along columns because of the strong dipole interaction along the columns, but need not be anti-ferromagnetic along rows because of the severely weakened dipole interaction along the rows.

Reference [14] considered an example of image restoration carried out with such a nanomagnetic array. Consider a white segment of an image where all pixels are white. Ideally every MTJ in the entire two-dimensional array should be in the high resistance state because every pixel is white. However, consider the situation when noise has corrupted the system, switching every MTJ along one particular row to the low resistance state, encoding black pixels. This situation is shown in the left panel of Fig. 10.3a. The system is not in the ground state since we do not have perfect ferromagnetic ordering along any row. Uniaxial stress of the appropriate sign (which will make the product of stress and magnetostriction negative) applied along the major axis of the nanomagnets (i.e., along the column), corrects the corrupted pixels and restores the original image, as shown in Fig. 10.3. If two rows, which are either adjacent or non-adjacent, are corrupted, that too can be corrected by applying stress. This is also shown in Fig. 10.3.

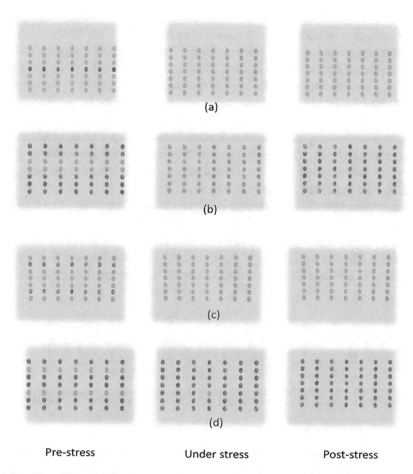

Pre-stress   Under stress   Post-stress

Fig. 10.3 **a** The white segment of an image (7 × 7 pixels) where the entire fourth row has been corrupted and turned black. For the sake of color contrast, we have used green = white; blue = black. Application and subsequent withdrawal of global stress of the appropriate sign corrects the corrupted row. **b** Array of black pixels in the black segment of an image where two consecutive rows have been corrupted and turned white. Stress application and withdrawal corrects both corrupted rows. **c** Array of white pixels in the white segment of an image where two nonconsecutive rows have been corrupted and turned black. Stressing corrects both rows. **d** Array of black pixels where two nonconsecutive rows have been corrupted and turned white. Again, stressing corrects both rows. This is an example of hardware based *image restoration*. Reprinted from *IEEE Trans. Elec. Dev.* **64** 2417 (2017) with permission of the Institute of Electrical and Electronics Engineers

Another example of straintronic image processing with a two-dimensional arrays of dipole coupled magnetostrictive nanomagnets is shown in Fig. 10.4. This is an example of "edge-enhancement-detection". A black-and-white image has a white segment and a black segment demarcated by a sharp boundary. Blurring occurs when the pixels get corrupted to result in a random patchwork of black and white pixels as shown in the left panel of Fig. 10.4. Application of global stress erodes the energy barriers in each nanomagnet to make dipole coupling set the state of each nanomagnet to correspond to the ground state configuration of the array. Ground state is restored by turning a black pixel white if its majority of neighbors are white and vice versa. This happens as a consequence of nearest neighbor dipole interaction between the nanomagnets and it enhances the edge contrast between the two segments, as shown in Fig. 10.4b.

The error probability in these tasks is negligible because the information is embedded in *many* devices rather than a single one. For example, in Fig. 10.4, the white segment is represented by 21 pixels, not just 1, and the black segment is represented by 28 pixels. Thus, there is inherent *redundancy* in the approach, since many devices together encode the information, and that gives it error resilience. It is a characteristic trait of collective

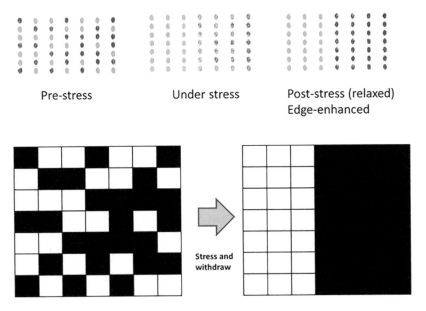

Fig. 10.4 Edge enhancement detection. The left segment (left three columns) has majority white pixels and the right segment (four right columns) has majority black segments. After application and removal of global stress applied along the major axes of the elliptical nanomagnets, the white-dominant segment turns all-white and the black-dominant segment turns all-black. This enhances the edge contrast between the two segments. Reprinted from *IEEE Trans. Elec. Dev.* **64** 2417 (2017) with permission of the Institute of Electrical and Electronics Engineers

computational hardware, regardless of whether it is implemented with straintronics or some other methodology. The purpose of this example was to emphasize that straintronics is ideally suited for this type of applications since it is forgiving of errors.

Hardware based image processing is attractive in many ways. It is superior to software based approaches because it is much faster (there is no need to execute instruction sets) and the processing time is almost independent of the image size (number of pixels) because all pixels are processed in parallel, not sequentially. The drawback, however, is that these processors are *application specific* since only very specific types of tasks can be executed with them. Neither are they universal processors, nor are they reconfigurable.

10.2   Straintronic Neuron/Perceptron

Neuromorphic computing and neural networks have emerged as a major computational paradigm for information processing since the early work of Little [15], Hopfield [16] and Mead [17]. It is the foundational basis of many aspects of artificial intelligence and machine learning. Its most popular uses today are in deep learning networks, spiking neural networks and reservoir computing. Neural networks benefit from low energy consumption (the human brain typically dissipates about 20 W to perform cognitive tasks) and hence it is a likely niche for straintronics.

A biological neuron takes inputs from other neurons via dendrites, processes that information within the soma, transmits the processed information through dendrites and feeds it to other neurons via synapses. The first computational model of a neuron was proposed by Warren McCulloch and Walter Pitts in 1943 [18]. In its simplest form, it receives a number of binary inputs x_i and aggregates them to create a sum $g(x_1, x_2 \ldots x_n) = g(\mathbf{x}) = \sum_{i=1}^{n} x_i$. The output y is a non-linear threshold function of $g(\mathbf{x})$:

$$y = f[g(\mathbf{x})] = \begin{cases} 1 \text{ if } g(\mathbf{x}) \geq \Theta \\ 0 \text{ if } g(\mathbf{x}) < \Theta \end{cases}, \text{ where } \Theta \text{ is a threshold. Here, the non-linear func-}$$

tion f is a threshold function. Other non-linear functions that can be used are sigmoid, piecewise linear, etc.

Later, inspired by the work of Donald Hebb [19], Frank Rosenblatt proposed the idea of the *perceptron* [20]. Its difference with the MCulloch-Pitts neuron is that it can take non-binary inputs and produce a binary output that is a non-linear (e.g. threshold function) of a *weighted* sum of the inputs $g(\mathbf{x}) = \sum_{i}^{n} w_i x_i + b$. Here, again, the output is a binary

variable $y = f[g(\mathbf{x})] = \begin{cases} 1 \text{ if } g(\mathbf{x}) \geq \Theta \\ 0 \text{ if } g(\mathbf{x}) < \Theta \end{cases}$. In its most advanced form, the weights are

learnable, meaning they can change depending on the history of the inputs received.

A magnetic McCulloch-Pitts neuron and/or a perceptron can be implemented with a magnetic tunnel junction (MTJ) whose resistance state can be switched abruptly from high to low, or vice versa, by passing a weighted (or unweighted) sum of input spin currents through the MTJ, which delivers a spin transfer torque or spin orbit torque on the soft layer [21–24]. Similarly, the weighted (or unweighted) sum of input voltages applied to a straintronic MTJ can make its resistance state switch from high to low, or vice versa, thereby implementing a straintronic neuron or perceptron [25].

Figure 10.5 depicts a straintronic perceptron implemented with a skewed straintronic MTJ identical to that in Fig. 10.1. The soft layer's magnetization has two stable orientations along $\Psi_\parallel$ and $\Psi_\perp$ because of the in-plane magnetic field **B**. The line joining the electrodes A and A' lies along $\Psi_\parallel$. The piezoelectric layer is poled vertically up. Because of any residual dipole coupling between the hard and soft layers, the MTJ will be in the high resistance state (soft layer's magnetization along $\Psi_\perp$) when there is no voltage input.

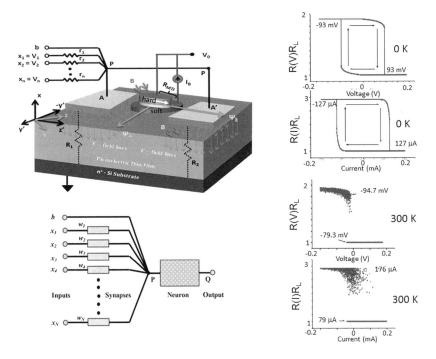

Fig. 10.5 (Left panel) A straintronic McCulloch-Pitts neuron or a perceptron implemented with a skewed straintronic magnetic tunnel junction and passive resistors acting as synaptic weights. The weights can be made programmable by replacing the resistors with memristors. (Right panel) The ratio of the MTJ resistance R(V) or R(I), at any voltage V appearing at node P or any current I passed through the MTJ, to the low resistance state R_L, as a function of V or I. The results are shown for 0 K temperature (no thermal noise) and 300 K temperature (thermal noise present). There is hysteresis because the device is magnetic. Reprinted from *Nanotechnology,* **26**, 285201 (2015) with permission of the Institute of Physics

Application of the input voltages of the right polarity generates biaxial strain underneath the soft layer—compressive along $\Psi_\parallel$ and tensile along $\Psi_\perp$ because of the placement of the electrodes A and A'. The magnetostriction of the soft layer is assumed to be negative. When the weighted sum of the voltages, plus the bias b, exceeds a threshold value where the stress generated can lower the energy barrier within the soft layer sufficiently, the soft layer's magnetization will switch to $\Psi_\parallel$ from $\Psi_\perp$, and the MTJ will switch suddenly to the low resistance state R_L.

The inputs x_i–s to the neuron and the fixed bias b are realized with voltages V_i and b; the latter is realized with a constant current source I $\left[b = I(R_1 \| R_2 \| r_1 \| r_2 \| \ldots \| r_{N-1} \| r_N) \right]$. The voltage appearing at node P in Fig. 10.5 is dropped across the piezoelectric layer underneath the (shorted) contact pads A and A'. This voltage is a weighted sum of input voltages and bias, and is given by (using standard superposition principle).

$$V_P = \sum_{i=1}^{N} w_i V_i + b,$$

where the synaptic weights are:

$$w_i = \frac{R_1 \| R_2 \| r_1 \| r_2 \| \ldots \| r_{i-1} \| r_{i+1} \| \ldots \| r_N}{R_1 \| R_2 \| r_1 \| r_2 \| \ldots \| r_{i-1} \| r_{i+1} \| \ldots \| r_N + r_i}.$$

The resistances R_1 and R_2 are the resistances of the piezoelectric layer underneath the contact pads and r_i–s are the series resistances (connected to the input terminals) that implement the programmable weights.

Reference [25] carried out stochastic Landau-Lifshitz-Gilbert simulations of magnetodynamics of the neuron firing and obtained the firing characteristics shown in the right panel of Fig. 10.5 at 0 K temperature and 300 K temperature. There is hysteresis since the device is magnetic. The scatter at room temperature is due to thermal noise, and thermal noise actually reduces the hysteresis in this case. Two devices were simulated: a straintronic neuron activated by voltage, and a spin transfer torque neuron activated by current. Based on these simulations, Ref. [25] concluded that the firing delay for the straintronic neuron is 0.9 ns and the energy dissipated to fire is 8 aJ, whereas for the current driven neuron, the firing delay is 6 ns and the energy dissipated to fire is 50 fJ, which is nearly four orders of magnitude larger. This showcases the energy efficiency of straintronics.

10.3   Straintronic Platforms for Solving Combinatorial Optimization Problems

Probabilistic computing is a powerful approach to solving a class of problems [26] that can benefit from using probabilistic bits (p-bits). P-bits have been termed a poor (wo)man's qubit. A classical binary bit is *either* 0 or 1 all the time, a qubit is a coherent superposition and hence *both* 0 and 1 all the time, whereas a p-bit is sometimes 0 and sometimes 1. At any given time, its value is 0 with a probability p and 1 with a probability $1-p$.

Recently there has been a burst of activity in implementing a probabilistic computer with nanomagnets whose internal energy barriers (due to shape or surface anisotropy) is low enough that thermal perturbations can randomly fluctuate their magnetizations [27–30]. They implement a binary stochastic neurons (BSN) which can either act as Ising spins in Ising computers for solving combinatorial optimization problems, or act as neurons in Boltzmann machines for solving optimization problems via an energy minimization approach [27]. The most convenient rendition of a BSN is a nanomagnet shaped like a circular disk as shown in Fig. 10.6. If the magnetization lies in the right segment, we will interpret that as encoding the bit 1 and if it lies in the left segment, we will interpret that as bit 0. Normally either bit will be equally probable, but if we inject a spin polarized current whose spin is aligned to the right, then bit 1 will be more probable whereas if we inject a spin current with spins aligned to the left, then bit 0 will be more probable. Thus, we can progam the probability with a spin current and this provides the basis for a binary stochastic neuron.

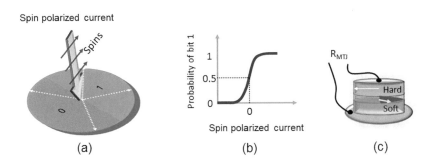

Fig. 10.6 **a** A circular nanomagnet whose magnetization fluctuates in time owing to thermal perturbations. If the magnetization is in the right semi-circle, it is interpreted as bit 1 and if it is in the left semi-circle, it is interpreted as bit 0. **b** The probability that the bit state is 1 as a function of a spin polarized current injected into the circular nanomagnet in a direction perpendicular to the plane. The spins are perpendicular to the current direction and aligned such that if one rotates the spins clockwise by 90°, they will align in the direction of the current. A magnetic tunnel junction is used to read the bit state. The hard layer is permanently magnetized to make its magnetization point to the left

The bit state at any instant of time is read with a magnetic tunnel junction (MTJ) whose hard layer is magnetized to the left. The MTJ resistance is $R_{MTJ} = R_P + \frac{R_{AP}-R_P}{2}[1 - \cos\theta]$, where θ is the angle between the magnetizations of the hard and soft layers. If the MTJ resistance is greater than $(R_{AP} + R_P)/2$, then it is interpreted as bit 1; otherwise, it is interpreted as bit 0.

Normally, the probabilities of 0 and 1 will be equal, but if a spin polarized current is injected into the MTJ, it will produce a spin transfer torque and make the probabilities unequal. Thus, with this construct, one can produce a binary state m_i (-1 or $+1$) [here we use the bipolar representation where -1 is interpreted as bit 0] at time step $(n+1)$ given by the analytical expression.

$$m_i(n + 1) = \text{sgn}[\tanh(I_i(n)) - r_i]$$

where I_i is the dimensionless input spin current that biases the output either towards -1 or towards +1 (depending on its sign) and r_i is a random number uniformly distributed between -1 and $+1$. Each BSN described by Equation (1) receives its input from a weighted sum of other BSNs obtained from a synapse $I_i(n) = \sum_j W_{ij} m_j(n)$. A wide variety of problems can be solved by properly designing or learning the weights W_{ij}, e.g. classification problems [31], constrain satisfaction problems [32], generation of cursive letters [33], etc.

To design a combinatorial optimization problem solver, we have to implement two ingredients: (1) biasing the BSNs by passing a spin polarized current through them, and (2) implementing programmable weights W_{ij} connecting nearest neighbor neurons. It is beneficial to implement the weights via *dipole coupling* between neighbors. Dipole coupling has many advantages: (i) it is "wireless" and consumes no area on a chip, (ii) there is no current flow involved and hence no I^2R loss, and (ii) dipole coupling effect can be modulated by straining both neighboring nanomagnets to lower or raise the energy barriers within them, thereby implementing synapses with *variable weights*. For dipole coupling to be effective in determining the magnetic states of a nanomagnet, it must be able to overcome the energy barrier within the nanomagnet. Hence, by modulating the energy barrier with strain, we can modulate the effectiveness of dipole coupling and hence the synaptic weight. Strain of one sign (tensile or compressive) will raise the energy barrier and strain of the opposite sign will lower it. Local strain is generated around a target synapse by applying a small voltage on the piezoelectric in the neighborhood of the synapse.

Many challenging optimization problems (e.g. graph connectivity problems—max-cut or min-cut) can be cast as an energy minimization problem. We write an energy (or cost) function as $E = [S]^T\{x\} + [S]^T[W][S]$ where $[S]$ is a $n \times 1$ column vector whose elements are 0 or 1 representing the binary states of n neurons, $\{x\}$ is a $1 \times n$ row vector representing the bias for the n neurons and $[W]$ is the $n \times n$ weight matrix W_{ij}. A problem is formulated by choosing $\{x\}$ and $[W]$, and then minimizing E to find the

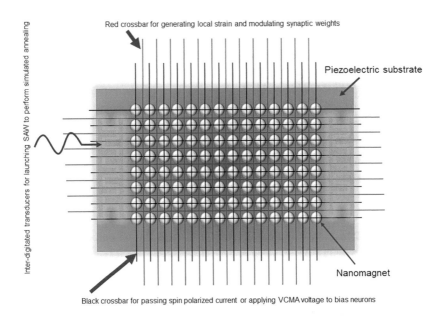

Fig. 10.7 Straintronic architecture for solving optimization problems with very low energy cost. BSNs are implemented with low barrier nanomagnets. One set of cross bars is used to bias any BSN by injecting spin polarized current, and the other set of crossbars applies strain to modulate any synaptic weight. The interdigitated transducers are used to launch a surface acoustic wave to perform simulated annealing to guarantee that the ground state has been reached

optimum solution. There are only two hardware requirements—a scheme to generate $\{x\}$ for biasing the BSNs and a scheme to program weights to generate $[W]$. They can be accomplished with two different sets of crossbars slightly displaced from each other as shown in Fig. 10.7. One set of crossbars is used to inject spin polarized current into any targeted nanomagnet (or BSN) to bias the probability that the output of the BSN is the bit 1(i.e. to program the vector $\{x\}$), and the other set of crossbars is used to generate strain around any targeted synapse to modulate its weight (i.e. to program the matrix $[W]$).

10.4 Straintronic Correlator/Anti-Correlator

For many computing applications in probabilistic computing, it is necessary to correlate or anti-correlate two p-bit streams [30]. Such a device (for generating correlation or anti-correlation) can be implemented with two elliptical dipole-coupled straintronic magnetic tunnel junctions (MTJs) [with in-plane magnetic anisotropy in the ferromagnetic layers] placed such that the line joining the centers of the two MTJs is either collinear with the minor axes, or the major axis. Their hard layers are magnetized parallel to each other. The former type implements an anti-correlator and the latter type a correlator. They are shown in Fig. 10.8. The distinguishing feature is that in the anti-correlator, dipole coupling would tend to make the magnetizations of the two soft layers mutually anti-parallel and therefore the resistance states of the two MTJs will be anti-correlated, while in the correlator, dipole coupling would like to align the magnetizations of the soft layers parallel to each other. This makes the resistance states of the two MTJs correlated.

One of the MTJs has much more shape anisotropy than the other, and is larger. The energy barrier within the former is very high. Hence, its magnetization state is not influenced by that of its neighbor's via dipole coupling even when stress is applied to lower the internal energy barrier by some amount. In the other MTJ with much lower internal energy barrier, stress can lower the energy barrier enough to allow the dipole influence of its neighbor to have some effect on its magnetization state. How much that effect is depends on the amount of stress applied since stress lowers the energy barrier in the smaller nanomagnet and hence varies the degree of influence that dipole coupling with the neighbor has on its magnetization state. Hence, by applying global stress, we can correlate or anti-correlate the resistance state (high or low) of the smaller MTJ to that of the larger MTJ to varying degrees.

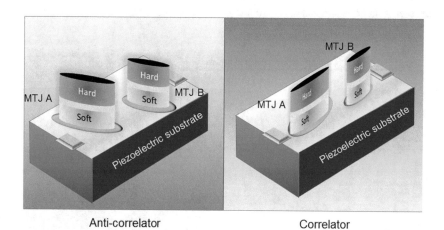

Anti-correlator   Correlator

Fig. 10.8 Two closely spaced magnetic tunnel junctions (MTJ)—one large with large shape anisotropy (MTJ A) and the other small with small shape anisotropy (MTJ B)—fabricated on a piezoelectric substrate. Because they are in close proximity, there is dipole interaction between them whose effect on the magnetization state in the smaller MTJ can be modulated with stress. The magnetization state in the larger MTJ is not influenced by dipole interaction with its neighbor because its internal energy barrier is large and stress cannot lower it sufficiently. This construct can act as a bit anti-correlator (left panel), and a bit correlator (right panel). Both can also act as a simple 2-node Bayesian network with the larger MTJ (with higher internal energy barrier) acting as the "parent" node and the other acting as the "child" node. To what extent the child's resistance state (high or low) is correlated or anti-correlated with that of its parent can be modulated with applied stress. The two electrodes shown are used for applying stress

The probability of the larger MTJ being in either resistance state is set independently by injecting a spin polarized current into it (see Fig. 10.6b). No spin polarized current is injected into the smaller MTJ and the probability of its being in either resistance state is correlated or anti-correlated with the resistance state of the larger MTJ.

In the anti-correlator, the anti-correlation between the two p-bit streams can be varied between 0% and 100% (meaning that at one extreme, one p-bit stream does not depend at all on the other, and at the other extreme, one p-bit stream is always the binary complement of the other) depending on the amount of stress applied to both MTJs simultaneously. Similarly, in the correlator, the correlation can be varied between 0% and 100% depending on the magnitude of global stress.

Figure 10.9a shows the degree of anti-correlation in a pair of the type shown in the left panel of Fig. 10.8 as a function of applied stress, reproduced from Ref. [34]. These results are calculated from coupled stochastic Landau-Lifshitz-Gilbert equation in the presence of room temperature thermal noise for various magnitudes of the spin polarized current that sets the p-bit in the larger MTJ. Here the value of 1 indicates no anti-correlation (or correlation), and the value of −1 indicates perfect (100%) anti-correlation. Note that the characteristic is non-monotonic and perfect anti-correlation is obtained (for large spin polarized current) not at the maximum stress, but at some intermediate value of the stress.

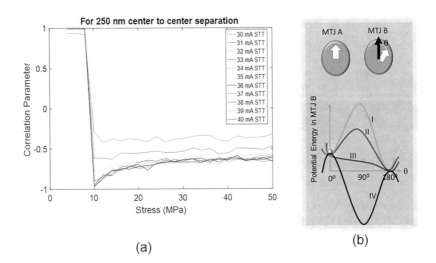

(a)   (b)

Fig. 10.9 a Degree of anti-correlation between the resistance states of two dipole coupled straintronic magnetic tunnel junctions versus stress for different spin polarized current injected into the larger MTJ. The value of +1 indicates no correlation or anti-correlation, whereas the value of −1 indicates perfect anti-correlation. b Potential profile within the soft layer of the smaller MTJ for different stress levels (I–IV). III corresponds to critical stress when the barrier is completely eroded but not inverted. Reproduced from *Sci. Rep.*, **10**, 12361 (2020) with permission from Nature Publishing Group

This stress is the *critical stress* where the effect of dipole coupling is maximal because the stress just erodes the potential barrier instead of inverting it, thereby allowing the magnetization in the soft layer of the smaller MTJ to reach the global ground state with near certainty.

10.5   Bayesian Inference Engines and Belief Networks

Bayesian (belief) networks are computational modules that are especially adept at computing in the presence of uncertainty (e.g. disease progression, stock market behavior, etc.). Its main use has been in predictive analysis; given a number of causes for a possible outcome, it can predict with high certainty which cause was actually responsible for the outcome. The simplest 2-node Bayesian network consists of a parent node and a child node, where the state of the "child" is influenced by the state of the "parent", but not the other way around. Bayesian networks are therefore referred to as "directed graphs" in computer science literature since influence is exerted unidirectionally from parent to child and not the other way around.

The two MTJs in Fig. 10.8 host p-bits and the larger one can be viewed as a binary stochastic neuron whose p-bit value (or the probability that the bit is 1) is set by a spin polarized current. Its resistance state is not influenced at all by the state of the smaller one. Hence, it can be viewed as the parent node. The smaller one is the child node whose p-bit value is influenced by the parent. If we view the dipole coupling as a synaptic connection between the two, then the synapse is clearly non-reciprocal since the parent influences the child, but not the other way around. This is very different from Boltzmann machines where the synapse is reciprocal.

It is also clear that the two-MTJ construct described here can act as a *conditional probability* encoder since we can set the probability that the smaller MTJ is in bit state 1 depending on the state of the larger MTJ. These features were exploited to construct a Bayesian belief network in Ref. [35]. Other uses of straintronic MTJs in implementing Bayesian inference engines have been described in Refs. [36, 37]. These networks benefit from the high degree of energy efficiency and the non-volatility of the probability generators [36, 37], which is why straintronics is ideal for their implementation.

References

1. F. Barahona, On the computational complexity of Ising spin glass models. J. Phys. Math. Gen. **15**, 3241 (1982)
2. M. Yamaoka, C. Yoshimura, M. Hayashi, T. Okuyama, H. Aoki, H. Mizuno, A 20 k-spin Ising chip to solve combinatorial optimization problems with CMOS annealing. IEEE J. Solid State Circuits **51**, 303 (2016)

3. K. Kim, M.S. Chang, S. Korenblit, R. Islam, E.E. Edwards, J.K. Fredericks, G.D. Lin, L.M. Duan, C. Monroe, Quantum simulation of frustrated Ising spins with trapped ions. Nature **465**, 590 (2010)

4. I. Mahboob, H. Okamoto, H. Yamaguchi, An electromechanical Ising Hamiltonian. Sci. Adv. **2**, e1600236 (2016)

5. S. Kirkpatrick, C.D. Gelatt, M.P. Vecchi, Optimization by simulated annealing. Science **220**, 671 (1983)

6. M.W. Johnson et al., Quantum annealing with manufactured spins. Nature **473**, 194 (2011)

7. L.O. Chua, L. Yang, Cellular neural networks: applications. IEEE Trans. Circuits Syst. **35** 1273 (1988)

8. S. Utsunomiya, K. Takata, Y. Yamamoto, Mapping of Ising models into injection-locked laser systems. Opt. Express **19**, 18091 (2011)

9. V.P. Roychowdhury, D.B. Janes, S. Bandyopadhyay, X. Wang, Collective computational activity in self-assembled arrays of quantum dots: A novel neuromorphic architecture for nanoelectronics. IEEE Trans. Elec. Dev. **43**, 1688 (1996)

10. K. Karahaliloglu, S. Balkir, S. Pramanik, S. Bandyopadhyay, A quantum dot image processor. IEEE Trans. Elec. Dev. **50**, 1610 (2003)

11. S. Bhanja, D.K. Karunaratna, R. Panchumarthy, S. Rajaram, S. Sarkar, Non-Boolean computing with nanomagnets for computer vision applications. Nat. Nanotechnol. **11**, 177 (2015)

12. B. Sutton, K.Y. Camsari, B. Behin-Aein, S. Datta, Intrinsic optimization using stochastic nanomagnets. Sci. Rep. **7**, 44370 (2016)

13. N. D'Souza, J. Atulasimha, S. Bandyopadhyay, An ultrafast image recovery and recognition system implemented with nanomagnets possessing biaxial magnetocrystalline anisotropy. IEEE Trans. Nanotechnol. **11**, 896 (2012)

14. M.A. Abeed, A.K. Biswas, M.M. Al-Rashid, J. Atulasimha, S. Bandyopadhyay, Image processing with dipole-coupled nanomagnets: Noise suppression and edge enhancement detection. IEEE Trans. Elec. Dev. **64**, 2417 (2017)

15. W.A. Little, The existence of persistent states in the brain. Math. Biosci. **19**, 101 (1974)

16. J.J. Hopfield, Neural networks and physical systems with emergent collective computational abilities. Proc. Natl. Acad. Sci. **79**, 2554 (1982)

17. C. Mead, *Analog VLSI and Neural Systems* (Addison-Wesley, Boston, 1989)

18. W.S. McCulloch, W. Pitts, A logical calculus of the ideas immanent in nervous activity. Bull. Math. Biophys. **5**, 115 (1943)

19. D.O. Hebb, *The Organization of Behavior* (Wiley, New York, 1949)

20. F. Rosenblatt, *Principles of Neurodynamics* (Spartan, New York, 1962)

21. M. Sharad, C. Augustine, G. Panagopoulos, K. Roy, Spin-based neuron model with domain-wall magnets as synapse. IEEE Trans. Nanotech. **11**, 843 (2012)

22. M. Sharad, D. Fan, K. Roy, Spin-neurons: A possible path to energy-efficient neuromorphic computers. J. Appl. Phys. **114**, 234906 (2013)

23. A. Sengupta, S.H. Choday, Y. Kim, K. Roy, Spin orbit torque based electronic neuron. Appl. Phys. Lett. **106**, 143701 (2015)

24. A. Sengupta, M. Parsa, B. Han, K. Roy, Probabilistic deep spiking neural systems enabled by magnetic tunnel junction. IEEE Trans. Elec. Dev. **63**, 2963 (2016)

25. A.K. Biswas, J. Atulasimha, S. Bandyopadhyay, The straintronic spin neuron. Nanotechnology **26**, 285201 (2015)

26. J. Pearl, *Probabilistic Reasoning in Intelligent Systems: Networks of Plausible Inference* (Morgan Kauffman, Burlington, Massachusetts, 2014)

27. K.Y. Camsari, B.M. Sutton, S. Datta, p-bits for probabilistic spin logic. Appl. Phys. Rev. **6**, 011305 (2019)

28. K.Y. Camsari, R. Faria, B.M. Sutton, S. Datta, Stochastic p-bits for invertible logic. Phys. Rev. X **7**, 031014 (2017)
29. B. Sutton, K.Y. Camsari, B. Behin-Aein, S. Datta, Intrinsic optimization using stochastic nano-magnets. Sci. Rep. 7, 44370 (2017)
30. W.A. Borders, A.Z. Pervaiz, S. Fukami, K.Y. Camsari, H. Ohno, S. Datta, Integer factorization using stochastic magnetic tunnel junctions. Nature **573**, 390 (2019)
31. U. Cilingiroglu, A purely capacitive synaptic matrix for fixed-weight neural networks. IEEE Trans. Circuits Syst. **38**, 210 (1991)
32. G.A. Fonseca Guerra, S.B. Furber, Using stochastic spiking neural networks on SpiNNaker to solve constrain satisfaction problems. Front. Neurosci. **11**, 714 (2017)
33. A. Mizrahi, T. Hirtzlin, A. Fukushima, H. Kubota, S. Yuasa, J. Grollier, D. Querlioz, Neural like computing with populations of superparamagnetic basis functions. Nat. Commun. **9**, 1533 (2018)
34. M. McCray, M.A. Abeed, S. Bandyopadhyay, Electrically programmable probabilistic bit anti-correlator on a nanomagnetic platform. Sci. Rep. **10**, 12361 (2020)
35. S. Nasrin, J. Drobitch, P. Shukla, T. Tulabandhula, S. Bandyopadhyay, A.R. Trivedi, Bayesian reasoning machine on a magneto-tunneling junction network. Nanotechnology **31**, 484001 (2020)
36. S. Khasanvis, M.Y. Li, M. Rahman, M. Salehi-Fashami, A.K. Biswas, J. Atulasimha, S. Bandyopadhyay, C.A. Moritz, Self-similar magneto-electric nano-circuit technology for probabilistic inference engines. IEEE Trans. Nanotechnol. **14**, 980–991 (2015)
37. S. Khasanvis, M.Y. Li, M. Rahman, A.K. Biswas, M. Salehi-Fashami, J. Atulasimha, S. Bandyopadhyay, C.A. Moritz, Architecting for causal intelligence at nanoscale. Computer **48**, 54–64 (2015)

Hybrid Straintronics and Magnonics

11

Magnonics is the science and technology of manipulating spin waves in magnetic systems using a variety of approaches. It has applications in information processing with spin wave logic [1] and analog computing [2], to name a few. Straintronics has enriched magnonics by providing a very convenient handle to tailor spin wave properties. In a hybrid strain-tronic and magnonic system, composed of magnetostrictive or magneto-elastic magnets fabricated on a piezoelectric film or substrate, the periodic strain generated by an acoustic wave will cause the magnetizations to precess because of the inverse magnetostriction (Villari) effect, thereby generating spin waves. The acoustic waves can also *amplify* and/or *modulate* the spin waves, or couple two different spin waves, thereby providing dynamic control over the spin waves.

When an array of magnetostrictive nanomagnets deposited on a piezoelectric substrate is placed in a magnetic field and an acoustic wave is launched in the substrate, at least *three types* of spin wave modes can be excited: (1) *Pure Kittel modes* caused by the precession of magnetization around the bias magnetic field. Their frequencies increase with the strength of the magnetic field in accordance with the Kittel formula [3]. They are pure spin wave modes uncoupled to the acoustic wave and will be present even if there was no acoustic wave, or if the magnets were not magnetostrictive and hence not subjected to magneto-elastic coupling. However, they will need the bias magnetic field to be present. (2) *Pure magneto-elastic modes* where the spin precession is caused by the periodically varying strain due to an acoustic wave (via the Villari effect). They can be observed without any bias magnetic field, but they require magnetostrictive magnets so that magneto-elastic coupling can take place. Their frequencies are obviously independent of the bias magnetic field (if such a field were present) since the latter is not required to generate them. (3) *Hybrid magneto-dynamical modes* [4] which are a mixture of the other two. They require the presence of a bias magnetic field as well as an acoustic wave and magneto-elastic coupling. Their frequencies do increase with the bias magnetic field, but obviously not in accordance with the Kittel formula.

11.1   Hybrid Magneto-Dynamical Modes

Spin wave modes excited by time-varying strain in a *single* magnetostrictive nanomagnet placed in a magnetic field were first reported in Ref. [4]. The modes were investigated with a two-color time-resolved magneto-optical Kerr effect (TR-MOKE) [5] measurement to extract the temporal magnetization oscillations in the nanomagnets. These oscillations were measured at different magnetic fields and then they were Fourier transformed to obtain the power spectrum at every magnetic field. Simulations were carried out to extract the power and phase profiles of these modes in the nanomagnet.

The sample consisted of an array of quasi-elliptical magnetostrictive Co nanomagnets deposited on a poled piezoelectric $Pb(Mg_{1/3}Nb_{2/3})O_3$-$PbTiO_3$ (PMN-PT) substrate. However, the probe beam of the TR-MOKE microscope was focused on a single nanomagnet, thereby interrogating only its spin wave modes. No surface acoustic wave (SAW) was launched externally, but such a wave existed in the substrate anyway. Two different mechanisms generated the SAW. First, the alternating electric field of the pump laser produced periodic compressive and tensile strain in the PMN-PT substrate from d_{33} and/or d_{31} coupling. The strain was tensile when the electric field in the substrate was in the same direction as the poling and compressive when the electric field was in the opposite direction. Second, the differential thermal expansions of the nanomagnet and the substrate underneath due to periodic heating by the pulsed pump beam also generated periodic strain. The combined effect of this periodic strain and the precession caused by the magnetic field led to the excitation of hybrid magneto-dynamical modes in the nanomagnet.

Figure 11.1 shows the time-resolved Kerr oscillations from the single nanomagnet reported in Ref. [5]. Three frequency components F_L, F and F_H (three peaks) were observed in the Fourier transformed power spectra. The central peak was the dominant peak (with the most power) at all bias magnetic fields except 700 Oe. The peak frequencies increased with increasing magnetic field, but not according to the Kittel formula, showing that they were hybrid magneto-dynamical modes (mixture of Kittel modes and magneto-elastic modes).

Simulated spin-wave mode profiles in the absence and the presence of a strain field revealed the spatial nature of the hybrid magneto-dynamical modes as shown in Fig. 11.2. The strain field's effect was simulated with an effective magnetic field H whose magnetostatic energy density was set equal to the strain anisotropy energy density according to $\mu_0 M_s H = (3/2)\lambda_s\sigma$. The strain σ was used as a fitting parameter to match theory and experiment as shown in Fig. 11.1. Instead of the characteristic center and edge mode behavior of a single nanomagnet excited optically or by pulsed magnetic field, the hybrid magneto-dynamical modes of frequencies F_L, F and F_H exhibit complex profiles with their unique characteristics, besides displaying rich variation with bias magnetic field, as seen in Fig. 11.2.

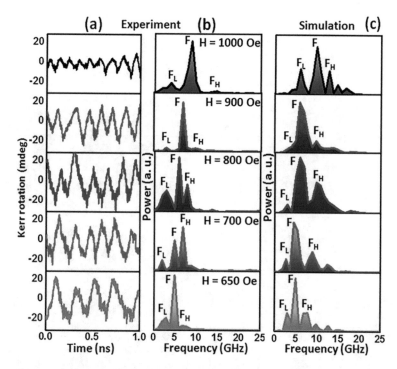

Fig. 11.1 Bias magnetic field dependence of the (background-subtracted) time-resolved Kerr oscillations from a single Co nanomagnet on a PMN-PT substrate. The pump fluence is 15 mJ/cm^2. **a** The measured Kerr oscillations in time and **b** the fast Fourier transforms of the oscillations, showing the power spectra. The spectral peaks shift to lower frequencies with decreasing bias magnetic field strength. There are multiple oscillation modes, each producing a peak. The dominant mode is denoted by F and its nearest modes are denoted by F_H and F_L (at all bias fields except 700 Oe where F_H is dominant over F). **c** Fourier transforms of the temporal evolution of the out-of-plane magnetization component at various bias magnetic fields simulated with the micromagnetic simulator MuMax3 where the amplitude of the periodically varying strain anisotropy energy density K_0 is assumed to be 22,500 J/m^3. The simulation has additional (weak) higher frequency peaks not observed in the experiment. The spectra in the two right panels are used to compare simulation with experiment. Reproduced from *ACS Appl. Mater. Interfaces*, **10**, 43,970 (2018) with permission of the American Chemical Society

11.2   Amplification of Spin Waves in a Two-Dimensional Periodic Array of Magnetostrictive Nanomagnets Fabricated on a Piezoelectric Substrate by a Surface Acoustic Wave

An array of two-dimensional magnetostrictive nanomagnets fabricated on a piezoelectric substrate, such as the one shown in the inset of Fig. 9.2, is sometimes referred to as a two-dimensional multiferroic "crystal" because the spatial periodicity mimics that of a

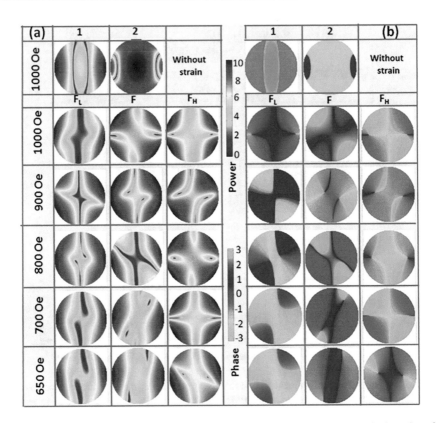

Fig. 11.2 Simulated **a** power and **b** phase profiles of the hybrid magneto-dynamical modes of frequencies F_L, F and F_H at different bias magnetic fields. These are resonant modes generated in the single (magnetostrictive) nanomagnet. The top most row shows edge and center modes at the two dominant frequencies in the Kerr oscillations in the absence of strain at 1000 Oe bias field and are provided only for comparison. The units of power and phase are dB and radians, respectively. Reproduced from *ACS Appl. Mater. Interfaces*, **10**, 43,970 (2018) with permission of the American Chemical Society

crystal. It is called a multiferroic system since the magnetostrictive/piezoelectric combination constitutes a two-phase multiferroic. In these systems, coupling of acoustic waves (launched in the substrate) with spin waves (in the nanomagnets) can transfer energy from the acoustic wave to the spin wave, thereby amplifying the latter.

Significant amplification would require significant energy transfer and hence a high degree of coupling. This would require "phase matching", meaning that the frequencies and wave vectors of the spin wave and acoustic wave will have to be about the same. That can happen only if the acoustic wave velocity is very close to the spin wave velocity.

Parametric amplification of spin waves in ferromagnetic thins films [6] and in magnonic crystals [7], via coupling to acoustic waves, has been demonstrated. In Ref.

[7], the spin wave power could be amplified by a factor of 7–8, even though the spin wave and acoustic wave were not phase matched. Spin wave amplification has important applications in spin wave logic and other spin wave devices since spin waves decay rapidly in ferromagnetic materials and will have to be amplified for logic level restoration.

11.3 Magnon-Phonon Interaction

The interaction between spin waves and ultrasonic waves was first studied theoretically by Charles Kittel in 1958 [8]. Experimental studies begun to emerge soon thereafter [9–15]. Interaction between spin waves and surface acoustic waves was studied in nanomagnet arrays by Yahagi et al. in 30-nm-thick nickel elliptic disks with varying array pitch (p) fabricated on a (100) silicon substrate with a 110-nm-thick hafnium oxide antireflection (AR) coating. A strong coupling between the two waves was observed when they were brought near degeneracy by varying an external magnetic field [16]. Three important manifestations of the magneto-elastic coupling that were observed were: (a) pinning of the spin-wave frequency over an extended range of the bias magnetic fields (which is characteristic of pure magneto-elastic modes), (b) generation of new mode with frequency differing by more than 100% from the intrinsic element response, and (c) sudden enhancement of the Fourier amplitude of the spin-wave mode at crossovers with acoustic modes since coupling becomes most efficient at these cross-over points (see Fig. 11.3). These features were explained by invoking an additional effective field component created by magneto-elastic coupling, and they were also accurately reproduced by simulations.

11.4 Strong Coupling Between Magnon and Phonon and the Formation of Magnon-Polaron

When magnons and phonons are strongly coupled, they form a hybridized magnon-phonon quasiparticle called a "magnon-polaron". In this state, the modes do not possess specifically either a magnon or a phonon character, but they co-exist in both states. Berk et al. reported the direct observation of coupled magnon-phonon dynamics in a single rectangular shaped Ni nanomagnet of dimensions $330 \times 330 \times 30$ nm grown on a (100) Si substrate capped by a 110-nm-thick hafnium oxide layer [16]. They utilized the vibrational modes of the single Ni nanostructure to stimulate phonon dynamics optically in the frequency range of 5 GHz–25 GHz along with the intrinsic precessional modes of the nanomagnet. The confined geometry of the Ni nanostructure created a confined cavity for the phonon and the magnon modes. By varying an external magnetic field in the appropriate geometries, the magnon mode was tuned through the phonon resonances to induce strong coupling manifested in the avoided crossing of the frequency-versus-magnetic-field branches [17]. Two different anti-crossings between two confined modes were observed,

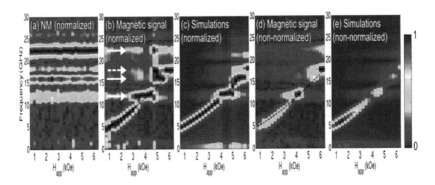

Fig. 11.3 Fourier spectra for nickel elliptic disks of $p = 212$ nm. **a** Measured frequency versus magnetic field of magneto-elastic modes, whose frequencies are field-independent. **b, d** Measured frequency versus magnetic field of Kittel or hybrid magneto-dynamical modes. Dashed line is the simulation result for nickel elliptic disks without magneto-elastic contribution. The solid and dashed arrows in (**b**) indicate the Rayleigh mode acoustic wave and surface skimming longitudinal acoustic wave frequencies. **c, e** Simulated magnetization dynamics including magneto-elastic coupling. The Fourier amplitudes in (**a**), (**b**), and (**c**) are normalized for better visualization of oscillation modes. The non-normalized Fourier spectra in (**d**) and (**e**) illustrate the enhanced Fourier amplitude at the crossover points at 12.2, 15.8, 17.7, and 22.3 GHz. Reprinted from Y. Yahagi, B. Harteneck, S. Cabrini and H. Schmidt, *Phys. Rev. B*, **90**, 140,405(R) (2014) with permission of the American Physical Society. © 2014 American Physical Society

and from the loss rate and coupling rate of those anti-crossings, the cooperativity factor was found to be about 1.14 and 0.74, which are indicative of intermediate coupling strength. This experiment showed that fairly strong coupling between acoustic waves and spin waves is possible in carefully engineered systems.

As mentioned earlier, when a magnon (spin wave) is strongly coupled with a phonon (acoustic wave), a new quasi-particle is born. It is called a *magnon-polaron*. The short lifetime of this particle often hinders the strong coupling required for its formation. This impasse was overcome by imposing spatial overlap of magnons and phonons in a Galfenol ($Fe_{0.81}Ga_{0.19}$) thin film nanograting (NG) grooved on an epitaxially grown Galfenol film on a (001)-GaAs substrate [18]. Galfenol, being a highly magnetostrictive alloy, possesses both enhanced magnon-phonon interaction and well-defined magnon resonances. The spatial overlap of the desired phonon and magnon modes resulted in strong coupling and revealed the presence of an optically excited magnon polaron. The authors showed that the symmetries of the localized magnon and phonon states play a decisive role in the magnon polaron formation and its ensuing manifestation in the optically excited magnetic transients (time-resolved Kerr rotation (KR)) measured by conventional time-resolved magneto-optical pump-probe spectroscopy. Figure 11.4 shows the hybridization of magnon and phonon modes.

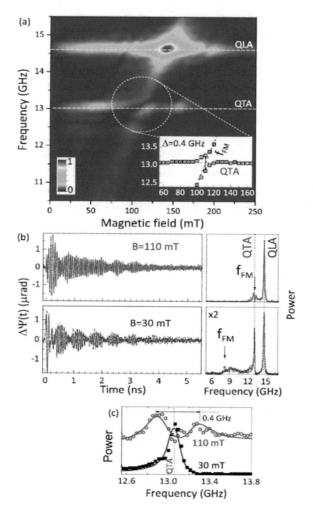

Fig. 11.4 Hybridization of magnon and phonon modes. **a** Color map which shows the spectral density of the measured Kerr rotation signal as a function of the external magnetic field applied along the NG diagonal when the interaction between the magnon and phonon modes of NG has maximal strength. The anticrossing is observed at $f = 13$ GHz and $B = 110$ mT. The inset shows the magnetic field dependence of the spectral peaks in the magnon spectrum around the intersection of the quasi-transverse-acoustic and ferromagnetic modes. **b** Transient Kerr rotation signals (left panels) and their fast Fourier transforms (right panels) at nonresonant ($B = 30$ mT) and resonant ($B = 110$ mT) conditions. Symbols show the measured signals and their fast Fourier transforms (FFTs); solid lines show respective fits and their FFTs. **c** Zoomed fragments of the FFT spectra shown in (**b**) around the resonance frequency. The splitting of the line in the resonance at $B = 110$ mT is clearly seen. Reprinted from F. Godejohann, A. V. Scherbakov, S. M. Kukhtaruk, A. N. Poddubny, D. D. Yaremkevich, M. Wang, A. Nadzeyka, D. R. Yakovlev, A. W. Rushforth, A. V. Akimov, and M. Bayer, *Phys. Rev. B*, **102**, 144,438 (2020) with permission of the American Physical Society. © 2020 American Physical Society

The theoretical possibility of generating magnon-polaron excitations via a spatially varying magnetic field was examined in Ref. [19]. The spatial variation results in strong magnon-phonon coupling. Such a coupling depends directly on the strength of the magnetic field gradient. The theory also predicted that control over coupling of specific phonon polarization to the magnons in the material can be obtained by tuning the direction of the magnetic field gradient [19].

Strong coupling between magnons and phonons has been exploited to measure magnon temperature [20] and it obviously has a role to play in spin caloritronics. It can also possibly find applications in quantum information science [21, 22]. So far, most experiments have shown that the coupling between magnons and phonons can be engineered with straintronic principles. Straintronics can play the role of an exquisite tool to fine tune and modulate the coupling between these two entities and this is one area where straintronics is now beginning to carve out a niche.

References

1. A. Khitun, M. Bao, K.L. Wang, Magnonic logic circuits. J. Phys. D: Appl. Phys. **43**, 264005 (2010)
2. A. Khitun, Magnonic holographic devices for special type data processing. J. Appl. Phys. **113**, 164503 (2013)
3. C. Kittel, On the theory of ferromagnetic resonance absorption. Phys. Rev. **73**, 155 (1948)
4. S. Mondal, M.A. Abeed, K. Dutta, A. De, S. Sahoo, A. Barman, S. Bandyopadhyay, Hybrid magneto-dynamical modes in a single magnetostrictive nanomagnet on a piezoelectric substrate arising from magnetoealstic modulation of precessional dynamics. ACS Appl. Mater. Interf. **10**, 43970 (2018)
5. A. Barman, A. Haldar, Time domain study of magnetization dynamics in magnetic thin films and micro- and nanostructures. Solid State Phys. **65**, Chapter 1 (2014)
6. I. Lisenkov, P. Dhagat, A. Jander, Inhomogeneous parametric pumping of spin waves by acoustic waves in an yttrium-iron-garnet film. https://ieeexplore.ieee.org/stamp/stamp.jsp?arnumber= 8007550
7. A. De, J.L. Drobitch, S. Majumder, S. Barman, S. Bandyopadhyay, A. Barman, Resonant amplification of intrinsic magnon modes and generation of new extrinsic modes in a two-dimensional of interacting multiferroic nanomagnets by surface acoustic waves. Nanoscale **13**, 10016 (2021)
8. C. Kittel, Interaction of spin waves and ultrasonic waves in ferromagnetic crystals. Phys. Rev. **110**, 836 (1958)
9. R.S. Silberglitt, Magnon-phonon coupling in metamagnetic systems: evidence for one phonon-two magnon interactions. AIP Conf. Proc. **24**, 230 (1975)
10. M. Foerster, F. Macia, N. Statuto, S. Finizio, A. Hernández- Mínguez, S. Lendínez, P. V. Santos, J. Fontcuberta, J. M. Manel Hernandez, M. Klaùi, L. Aballe, Direct imaging of delayed magneto-dynamic modes induced by surface acoustic waves. Nat. Commun. **8**, No. 407 (2017)
11. M. Bombeck, A.S. Salasyuk, B.A. Glavin, A.V. Scherbakov, C. Brüggemann, D.R. Yakovlev, V.F. Sapega, X. Liu, J.K. Furdyna, A.V. Akimov, M. Bayer, Excitation of spin waves in ferromagnetic (Ga, Mn)As layers by picosecond strain pulses. Phys. Rev. B **85**, 195324 (2012)

12. A.V. Scherbakov, A.S. Salasyuk, A.V. Akimov, X. Liu, M. Bombeck, C. Brüggemann, D.R. Yakovlev, V.F. Sapega, J.K. Furdyna, M. Bayer, Coherent magnetization precession in ferromagnetic (Ga, Mns)As induced by picosecond acoustic pulses. Phys. Rev. Lett. **105**, 117204 (2010)

13. L. Thevenard, I.S. Camara, S. Majrab, M. Bernard, P. Rovillain, A. Lemaître, C. Gourdon, J.-Y Duquesne, Precessional magnetization switching by a surface acoustic wave. Phys. Rev. B **93**, 134430 (2016)

14. M. Weiler, L. Dreher, C. Heeg, H. Huebl, R. Gross, M.S. Brandt, S.T.B. Goennenwein, Electrically driven ferromagnetic resonance in nickel thin films. Phys. Rev. Lett. **106**, 117601 (2011)

15. J. Janusonis, C.L. Chang, P.H.M. van Loosdrecht, R.I. Tobey, Frequency tunable surface magneto elastic waves. Appl. Phys. Lett. **106**, 181601 (2015)

16. Y. Yahagi, B. Harteneck, S. Cabrini, H. Schmidt, Controlling nanomagnet magnetization dynamics via magnetoelastic coupling. Phys. Rev. B **90**, 140405(R) (2014)

17. C. Berk, M. Jaris, W. Yang, S. Dhuey, S. Cabrini, H. Schmidt, Strongly coupled magnon-phonon dynamics in a single nanomagnet. Nat. Commun. **10**, 2652 (2019)

18. F. Godejohann, A.V. Scherbakov, S.M. Kukhtaruk, A.N. Poddubny, D.D. Yaremkevich, M. Wang, A. Nadzeyka, D.R. Yakovlev, A.W. Rushforth, A.V. Akimov, M. Bayer, Magnon-polaron formed by selectively coupled coherent magnon and phonon modes of a surface patterned ferromagnet. Phys. Rev. B **102**, 144438 (2020)

19. N. Vidal-Silva, E. Aguilera, A. Roldán-Molina, R.A. Duine, A.S. Nunez, Magnon-polarons induced by a magnetic field gradient. Phys. Rev. B **102**, 104411 (2020)

20. A. Agrawal, V.I. Vasyuchka, A.A. Serga, A.D. Karenowska, G.A. Melkov, B. Hillebrands, Direct measurement of magnon temperature: New insight into magnon-phonon coupling in magnetic insulators. Phys. Rev. Lett. **111**, 107204 (2013)

21. J. Li, S.Y. Zhu, Entangling two magnon modes via magnetostrictive interaction. New J. Phys. **21**, 085001 (2019)

22. H.Y. Yuan, Y. Cao, A. Kamra, R.A. Duine, P. Yan, Quantum magnonics: When magnon spintronics meets quantum information science. Phys. Rep. **965**, 1 (2022)

Printed in the United States
by Baker & Taylor Publisher Services